Attack of the Genetically Engineered Tomatoes

Hope you enjoy it.

Nicola Hawth
14/3/98

Peter Birch

Attack of the Genetically Engineered Tomatoes

The ethical dilemma of the '90s

Nicola Hamilton

Illustrated by Peter Birch

Whittet Books/Nemesis Press

First published in 1998 by Whittet Books Ltd./Nemesis Press, Hill Farm, Stonham Road, Cotton, Stowmarket, Suffolk, IP14 4RQ, England.

Copyright © 1998 by Nicola Hamilton
Illustrations © 1998 by Peter Birch

All rights reserved. No part of this publication may be reproduced, transmitted, or stored in a retrieval system, in any form or by any means, without the prior permission in writing from the publishers.

The right of Nicola Hamilton to be identified as the author of this work has been asserted in accordance with the Copyright, Designs and Patents Act 1988.

British Library Cataloguing in Publication Data. A catalogue record for this book is available from the British Library.

ISBN 1 873580 39 8

Printed and bound by WBC Book Manufacturers.

For nana and grandad

Contents

Acknowledgments		9
Introduction		11
Chapter 1:	Miracles in Action	21
Chapter 2:	This is Your Life	31
Chapter 3:	Attack of the Genetically Engineered Tomatoes	47
Chapter 4:	Do Shepherds Dream of Transgenic Sheep?	65
Chapter 5:	Ignorance is Bliss?	81
Chapter 6:	Killer in Us All	97
Chapter 7:	The Transformative Years	111
Epilogue		127
Bibliography		131
Index		141

Acknowledgments

Firstly, thank you Peter for nine years of fun, enthusiasm and limitless belief. Oh yes, and the great pictures.

I thank my mum and dad, and all my friends and family for their undying interest, support and encouragement.

Special thanks also to Gary, Jason, Brenda and Barry Stevens, and all the family, for their enormous warmth and friendliness, and for teaching me a great deal.

Thanks to my brother, James, for his suggested edits and revisions.

I also extend my deepest appreciation of the technical and human support of Cobra Systems & Programming Ltd., Steve Cabot of Haybridge Services Ltd., and Mike Palmer.

Finally, 'gark' to the penguins and Ranger.

Except where specifically acknowledged, much of the information included in this work is believed to be 'common knowledge' and its source is many and varied. Whilst there has been no verbatim use of copy, it is possible that some information has been gleaned from popular science press publications such as *New Scientist* and this is gratefully acknowledged.

INTRODUCTION

Birth.
Life.
Death.
Three small words that encompass your lifetime - your entire existence on Earth. But three huge concepts which you take for granted most of the time.

Out of all the people you could have been, fate chose the cells, the genes, that made you. Out of all of the hundreds of thousands of possible people that could have been created that night your parents had sex, you were the lucky one. And the world will never see those other individuals. It will only know you. For twins, this special concept is more easily grasped - they have shared this honour with each other, and they have grown and developed together in the womb from the first cells that were their new lives. For the rest of us, an appreciation of our origins, our design and uniqueness, requires a little more thought, or even a major detour from our often shallow, hedonistic journey through life.

The dice that destiny rolls when it creates new life are multi-sided - not just 10-sided or 20-sided, but almost *infinity*-sided. This illimitable scope for variety is a direct result of the beautifully designed molecule that is the blueprint for our body - DNA. The DNA molecule can be imagined as a twisted rope-ladder; it consists of 2 long strands, linked together at intervals by 'rungs', and twisted in a spiral shape known as a double helix. The ladder is immensely long, with thousands of rungs down its length, although much of it is usually kept coiled up for safe keeping as it is not needed.

It is these very long rope-ladders of DNA that we call chromosomes. Within most of the cells of our bodies lie 46 chromosomes, and each set is an identical copy of the very first 46 chromosomes that began our lives. As we grew and developed, each new cell was given a copy - the set of DNA that defines our biological soul. In 1997, researchers in the US managed to construct for the first time ever an artificial human chromosome. DNA was taken from white blood cells of the laboratory staff and fashioned into a new chromosome, with segments at both ends to stop the DNA from unravelling and a proper central region to give it structural support - just like a normal chromosome. But did they create life? Technically, yes - they constructed a model which mirrors that in nature and which can be put to use in helping to understand how chromosomes work. But practically, no - the messages contained in chromosomes are complex and interact with one another. A mock-up that is thrown together and exists as a separate entity cannot be considered a working model for new life.

There is a mysterious relationship among our chromosomes, from which there emanates an unknown force that permeates the rest of the cell. They seem to be able to communicate with one another, organising themselves and moving into the correct position when required. Part of this strange ballet stems from a curious liaison. The set of 46 chromosomes naturally falls into 23 pairs, each member having a strange affinity for the other. They also look similar, are of roughly the same size and always go together. (This makes it easy for researchers who have labelled the pairs 'Chromosome 1', 'Chromosome 2' etc.) When the time comes for the cell to divide, the internal machinery rolls into action and all the chromosomes sort themselves out. One member of each of these pairs has been passed down from our mother and one from our father, which helps to explain why we tend to show characteristics of both. Which member of the pair you inherit from each parent is entirely random; it is extremely unlikely that you and a sibling will inherit exactly the same 23 chromosome pairs from both of your parents, and this partly explains why we get variety within families as well as in the population as a whole.

The nature of one of these pairs is very special - because it determines our gender. The 'sex chromosomes' pair are termed 'XX' (females) or 'XY' (males). This means that women have 2 X chromosomes (so-called because of their normal 'X' shape which is similar to all of the other chromosomes), and men have 1 X chromosome and 1 Y chromosome (the latter being of unusually stunted appearance). Because there is an equal chance of inheriting the X or Y chromosome from your father, nature has ensured a roughly 50:50 split of men and women in the world. Interestingly, one of the X chromosomes that a woman inherits is not usually active. At a very early stage in the development of a female fetus, one of the pair is randomly chosen to be inactivated, and is packaged for storage. If this does not happen, there may be some quite unpleasant effects such as mental retardation and skeletal problems.

Inheriting the correct number of chromosomes is also important. If you gain an extra chromosome by accident it is usually a bad thing. For example, an extra Chromosome 21 results in Down's syndrome - a condition which is typically associated with reduced mental capacities, among other problems, and many children die before the age of 10. Extra sex chromosomes can also be inherited with undesirable results. About 1 in 1,000 men have an extra X chromosome, so they are 'XXY' - a condition known as Klinefelter's syndrome. Boys with this condition appear as normal until around puberty and then they begin to acquire more female features and become underdeveloped in their genitalia, as the conflicting messages of the extra X chromosome become apparent. Conversely, some males who inherit an extra Y chromosome appear to be more aggressive and may even have a greater tendency towards criminal behaviour. In the 1960s, in a number of countries, several enterprising lawyers realised the potential of this and instructed defendants in murder trials, for example, to plead a lack of criminal responsibility due to their 'XYY' chromosomes. Some of those tried actually did have their sentences reduced or quashed as a result of their genetic abnormality.

No one understands completely why having extra DNA can be detrimental. We do know, however, how our DNA normally works - and it is a marvellously simple process. When the chromosome needs to send a message to the rest of the cell - to order it to manufacture a particular enzyme for instance - it has to pull apart slightly, to expose a certain string of codes. The two long strands of the rope-ladder structure keep the DNA stable, but the rungs can break apart down the middle. This splitting apart is often called 'unzipping' for obvious reasons. After unzipping, the 'half-rungs' of the ladder lie exposed, and it is these half-rungs that hold the important information - the genetic code. Although all humans usually carry very similar sets of the 46 chromosomes - similar enough to make us all look human - we all have slight differences in our genetic codes, which make us all different people.

But how does it work? How do we make the astronomical leap from microscopic molecules in cells, which carry a kind of 'code', to a full-size, living, thinking, human being? The real answer is that nobody actually knows. But we can look at how individual lengths of DNA work on a day-to-day level, and use our imaginations to extrapolate beyond. The length of DNA that forms a chromosome is divided into thousands of sections of code, each responsible for programming a particular job. A section of DNA code that orders a specific message is called a gene. We often say phrases like 'genes for green eyes' meaning that there are particular genes which give people green eyes, or 'it's in the genes', meaning that there is some feature that a person must have inherited. Ultimately, all your features, or characteristics, have been created by the messages from your genes. Most of the time, a number of different genes act together to produce an effect in an individual, but occasionally one gene is of singular, vital importance. For example, in 1991, a startling piece of research discovered that just a single human gene on the Y chromosome, called *SRY*, is responsible for making a man a man. Researchers in London who were experimenting with mice found that by transferring the corresponding gene in mice, called *Sry*, into female embryos, the previously female mice developed into males.

One of the most incredible features of DNA is that it is responsible for programming the development and life of not only humans, but all other animals and plant life, and even bacteria. And yet there are just 4 different types of 'half-rungs' which are present in *any* piece of DNA. It seems unbelievable that four 'letters' of code are sufficient to construct the language for such a plethora of shapes and forms, but it is, in fact, all that nature has required. The trick lies in the combination of these four letters; a single gene may consist of thousands of letters, and these can appear in any order.

When the DNA ladder has unzipped along the stretch of a gene, the code lies exposed and can be 'read' by other molecules in the cell. There is a highly efficient team of workers on standby at all times, and they work quickly to carry out orders as soon as the DNA unzips. Messengers read the code by making a mirror image of it, and then move to the manufacturing facility of the cell to transfer the message to the engineers. Each message results in the production of a certain kind of protein - molecules that are crucial for most of the structure and function of our bodies. From the enzymes that help digest our food, to the hormones that regulate our blood sugar levels, and to the proteins of our cell walls that give us form - all are manufactured on the orders of our genes.

Over the last 20 years or so, scientists have been eagerly analysing the entire set of human chromosomes, called the *genome*, attempting to work out the nature of all of our different genes. The most obvious benefit of knowing about our genes is to help cure genetic diseases, but many other areas are being researched - especially when there's big money involved for those who find the solutions. For example, what would we give for a cure for baldness or greying hair? Japanese researchers have recently discovered that the production of natural pigments that give us hair colour is dependent on a certain kind of enzyme called 'tyrosinase'. If they could manufacture the gene that is responsible for producing tyrosinase and insert it into the hair follicles of greying temples, the cells might be able to produce the pigment for hair colour again. In California, researchers have already demonstrated an effective method of getting DNA into hair follicles by fitting it into loops of bacterial DNA, called 'plasmids', and packaging it in fat droplets. Once they have identified the gene that can promote hair regrowth, they hope to simply insert it into the hair follicles to help restore a new head of hair.

Even a genetic cure for ageing is being investigated in the US - could gene therapy be the ultimate cure for the most inescapable of human conditions? An enzyme called 'collagenase' breaks down proteins that are needed to protect against wrinkles. Older skin cells tend to produce more collagenase, and hence we tend to get more and more wrinkles throughout life. If researchers could find a way to stop the genes in cells from producing collagenase, could we protect our skin from wrinkles?

Part of the problem with research into ageing is that no one is entirely sure what causes it. One theory suggests that it is the accumulation of random damage to the DNA itself which occurs throughout our lives. Another suggests that ageing is simply due to a fatigued body, which becomes increasingly unable to produce new healthy cells. Fresh cells have to be prepared constantly by our bodies to replace old and worn-out ones. For each new replacement cell, all of the DNA in the chromosomes of the old cell has to be unwound and unzipped, in order that a complete replica can be made. Studies into ageing have recently focused on insights gained from Werner's syndrome - a rare but cruel disease which causes such premature ageing that those affected look very old, even in their twenties. People with this condition have inherited two defective copies of a gene needed to produce helicase, an enzyme that unwinds DNA, allowing it to be 'read' by the rest of the cell. It is thought that inserting extra copies of the helicase gene into cells may help to encourage healthy cell division, and hence slow down the ageing process.

In addition to normal cell division which goes on in all parts of our bodies all the time, there is a special kind of division which only occurs in the testes and ovaries, to produce sperm and egg cells. In this unique process, the original chromosome number of 46 has to be halved. As part of the preliminary manoeuvring, each pair of chromosomes indulges in their very intimate liaison - so intimate, in fact, that the two members may touch each other along their lengths. In doing so, certain genes are swapped between them in a process called 'recombination'. This means that for every egg or sperm cell that is produced, the chromosomes all carry different combinations of genes - and hence possible characteristics. Some characteristics that may have been hidden in the parent, masked perhaps by the effects of another gene, may reappear in the next generation. Even entirely new effects may be seen, as a different set of genes work together.

One of the most peculiar findings regarding our genes was stumbled upon in the 1980s. Researchers found that some genes behave differently depending on whether they are inherited from our mother or father. These are called 'imprinted genes', as there is obviously some kind of imprint left from the original parent which exerts an effect in the offspring. It now appears that in some cases, the copy of an imprinted gene inherited from your mother tends to override the paternal copy, and vice versa. For people who mistakenly get both chromosomes of a pair from just one parent (i.e. they don't receive the imprinted genes necessary from the other parent), there can be serious consequences. For example, some cases of the genetic obesity condition known as Prader-Willi syndrome and even childhood cancers have been linked to this problem.

The most fascinating research into imprinted genes concerns intelligence. Work carried out on mice suggests that the mother has more of an input into intelligence than the father. It appears that *maternal* genes are responsible for the development of the fetus and baby, and contribute substantially towards brain development and intelligence. And the *paternal* copies of genes express themselves for the development of the placenta rather than the fetus, and are more concerned with the emotional aspects of its character, and instinctive behaviours such as fighting and reproduction. Mice embryos which were manipulated to receive extra maternal genes instead of paternal ones grew into fetuses with huge heads and brains on small bodies. Vice versa, embryos which were given extra paternal genes developed huge bodies and very small brains. To ensure the optimal evolution of the human species, it would seem that men should look for high intelligence in their wives, and women should seek well adjusted, happy men!

Introduction

SCIENTISTS HAVE RECENTLY DISCOVERED GENES
FOR A VARIETY OF BEHAVIOURS.

Physical characteristics that we can see, like eye colour and height, are easily thought of as inheritable. But more abstract qualities, such as intelligence and behaviour, are less easily attributable to our genes. And there is much debate over the relative importance of our DNA and external, environmental influences - the nature versus nurture speculation. However, there is a substantial body of evidence which does suggest that genes have got a role to play in many of these characteristics. Research in twins, for example, has shown that genetic differences may account for up to 60 to 70% of variation in intelligence. The tendency towards aggression has also been studied in man as well as other animals, and at least one inheritable factor has been pinpointed; the enzyme, monoamine oxidase A, controls various levels of chemicals in the body and brain, necessary to regulate aggressive behaviour. In experiments with mice, it was found that animals that lacked the gene to make this enzyme tended towards violence and sexual aggression. Researchers in The Netherlands who investigated a Dutch family with a history of violence also found evidence that a defect in the monoamine oxidase A gene had an effect.

In a similar study, mice that lacked the gene for generating nitric oxide were found to have aggressive, antisocial tendencies; if caged in groups overnight, at least one or two individuals would be found dead in the morning. In addition, when placed in a cage with females, mice lacking the nitric oxide gene attempted to mate with them more frequently than the normal mice did, and they ignored the protests of the females. This would appear to be the worst end of the spectrum of violence, but a number of levels of antisocial behaviour have been investigated - in humans as well. For example, researchers at the University of Wales studied over 80 pairs of adolescent twins, and found that various types of antisocial behaviour, such as getting drunk and fighting, were to a certain extent genetically inherited! Even alcoholism and the related withdrawal symptoms have been linked to genetic make-up.

Physically and mentally, it seems that more and more aspects of our bodies can be traced to particular genes. And so it becomes increasingly important for us to take care of ourselves - genetically. When our DNA is working well and we experience no problems with our bodies, we take it for granted and tend not to worry about it. But we must be aware that it can be damaged, or *mutated*, and that the consequences may be permanent. There are lots of potentially harmful substances that can have an adverse effect on our DNA - various chemicals, drugs, ultraviolet radiation and X-rays, for example, are known *mutagens*. In the 1940s, the first demonstration of chemical mutagenesis was made with mustard gases that had been used in gas warfare - results which were, in fact, kept from publication by the military for a number of years. And we now know that everyone is exposed to various mutagens every day. Some estimates suggest 10,000 points of damage occur in our DNA every day. Japanese researchers recently proposed that the must powerful mutagen is a compound called 3-nitrobenzanthrone, a substance emitted in diesel exhaust fumes.

Fortunately, there is a proof reading and repair mechanism in our cells which is able to correct errors in our DNA. However, if the bombardment by unpleasant effects is too great, the correction mechanism cannot cope, and the mutations that occur will then persist throughout our lifetime. If the mutations are present in our reproductive cells, they will be passed down through future generations, affecting not only our children, but our children's children *ad infinitum*. Ten years on from the 1986 Chernobyl nuclear power plant disaster in Russia, a study was conducted to test for the effect of radioactive fallout on DNA mutations. Blood was collected from 79 families in three of the worst affected towns in Belarus and compared with that from 105 unaffected families in Britain. The children who were born to parents exposed to radioactive fallout were found to have twice as many mutations as the British children. This suggests that permanent damage was created in the cells of the Russian families, which will now be passed on through all future generations. And humans are not the only ones affected. The mutation levels in DNA of voles in a similar area was also found to be significantly above average. Who knows what other species have suffered in the way of mutations?

It is not yet known what the long term health consequences will be for the children of Chernobyl. One thing we do know for certain is that the vast majority of DNA mutations are generally bad for you. Even minute changes in your DNA can have dire consequences for your body if they occur in the wrong place. But ironically perhaps, it is in the mutation of DNA that the solution to the ultimate mystery could lie - the history of our species.

It is one of the most intriguing features of DNA that this simple molecule may hold the key to our evolution - and that of all living things on Earth. But only by randomly mutating, i.e. *going wrong*, can it have done this. Although the vast majority of mutations are deleterious, very occasionally one may occur that produces a useful or enhanced characteristic in an individual. In theory, any organism that finds itself better equipped will survive more easily than its compatriots, and gradually the beneficial change will work its way through the population to become a standard characteristic.

One of the main foundations of this traditional evolutionary argument is that the beneficial changes in DNA do not happen as a direct response to some outside, environmental pressure; giraffes, for example, did not acquire long necks *because* they had to reach high leaves, but because chance DNA mutations occurred which *fortuitously* gave them long necks. Hence they were more likely to survive and they eventually replaced giraffes with shorter necks.

But this may be too simple a theory. We know that DNA is a very smart molecule and there are undoubtedly many more things we have to learn about it. It must be guided by an intelligence of a sort to be able to programme such a complex creature as man, with the correct development, alignment and day-to-day working of millions of different cells - altogether forming a network of organs, transport systems and communication lines.

And it even has a truly selfless quality; DNA is quite happy to put the good of the rest of the body above its own existence. It accepts when its time has come and does the decent thing by committing suicide, in a process known as 'apoptosis'. It is a final altruistic performance of monumental proportions - seemingly in contradiction to the struggle for life which must have driven it from its first inception. For the greater good of the body, it not only orders the calm disassembly of the rest of its cell, but it also arranges its own execution - being neatly chopped up into thousands of segments.

Ultimately, DNA can be considered the fundamental basis of our biological existence - our birth, life and death. To know it is to appreciate its importance. And the implications of uninvited interventions of genetic manipulation by society.

1

MIRACLES IN ACTION

Let's start at The Beginning.

If we are to believe the cosmologists, it all began around 15 billion years ago in a violent rage of hydrogen and helium. Calm oblivion became cosmic inferno. Spontaneously created matter spewed forth into an unsuspecting dimension. The peaceful void of nothingness had collapsed and in its place was a loud new reality. The World had begun.

The theory of The Beginning is often explained in galactical jargon, by earnest scientists with flamboyant hand gestures. Curious phrases such as 'Hubble's constant' and 'the redshift effect' are thrown into the conversation with little explanation. But there is one term which has caught the imagination of layman and expert alike, and allows us to express in a simple fashion, the unimaginable - the 'Big Bang'.

Either by some incredible deal by fate or by a plan of supreme engineering, did the entire fabric of space and time come into existence. Since then, it has been continuously expanding, pushing its boundaries further and further into an empty black ether. And this mass, which we humans rather tamely refer to as the Universe, is now so hideously immense that we can have no hope of grasping its enormity.

Within its depths, sitting quite innocuously amidst the vastness of it all, is our own insignificant little planet called Earth. It is, as far as we know, the product of rampaging, stupendous forces which careered through our solar system, miraculously shaping a few raw ingredients of space dust and meteorites. A mere whipper snapper of a planet compared to some, it is currently celebrating its 4½ billionth birthday (give or take a few years).

The Big Bang was certainly one of those momentous occasions worthy of a mention in the chronicles of time. But an even more incredulous event was yet to come. Not content with the formation of matter and a few tricky conundrums to test modern physics students, *nature* soon began to crave more exciting projects - something to really stretch her skills. Consequently, a mere one billion years after the Earth had formed, a few organic molecules were miraculously arranged into microscopic single-celled organisms. Miniature units of life, these unique creations could not only survive from one day to the next but they could also reproduce, creating new generations to perpetuate the cycle.

Life on Earth had begun.

Three and a half billion years later, the Universe finds you sitting in your armchair, reading a book describing the amazing concept of your existence. Not only do your eyes read the words, your muscles turn the pages and your heart beat faster with excitement, but your brain manages to absorb the information and even allows you to form your own ideas.

From out of the great swathe of nothingness which occupied this space 15 billion years ago, we have intelligent life on Earth.

It's life Jim, but not as we know it

If we find it hard to comprehend the full scale of the Universe and its creation, how can we even begin to comprehend the true magnificence of *life* and just how ridiculously improbable the whole business is? It's not just the existence of living, thinking creatures like ourselves that poses the problems - but it's also the fact that a whole spectrum of weird and wonderful plants and animals have somehow got here. And millions of even more peculiar fantasies of nature have arisen and disappeared throughout the Earth's history, when we weren't around to notice.

From our own individual, blinkered perspectives, *life* is a limited time span, filled with artificial daily routines. We're born and suffer schooling to adulthood. Then the years roll by - washing the car on Sunday, Chinese takeaway on Monday, Majorca in the summer. And maybe we reproduce along the way, creating a few more individuals to carry on the cycle. We grow old by the fire, or in the garden. Then we die, from old age if we're lucky, or from one of a selection of unpleasant afflictions.

Perhaps this is a rather dismal view of what *life* has come to mean for many of us, in Western nations at least. But how often do you hear yourself moaning about a chore, a colleague, the weather or the state of the Government? Compare this with how frequently you revel in the miracles of life that have transformed the surface of the Earth, and those which take place throughout our lives - the single cell that becomes a fully formed, thinking human being... the existence and daily struggles of tens of millions of species in every niche of the planet... the intricate functioning of each of our bodies from one day to the next...

To gain a new perspective, let's look in the mirror and pause for thought. Who is it that stares back out? *What* is it? We are a whole being, but composed of many working parts - an eye, a muscle, a bone, a brain. And each of these working parts is composed itself of many millions of tiny compartments.

It is these imperceptible, beautifully tuned units that form the basis of all life on Earth, and they have performed their task astoundingly well for a good few billion years. With his usual style of supreme understatement, man has seen fit to name these units 'cells'. And by an unfortunate coincidence of language, the word has also come to mean 'a small room for a prisoner', 'a hermit's one-roomed dwelling' and 'a vessel containing electrodes for current-generation'!

The beauty of a cell is two-fold. The first thing to imagine is that each cell is a living, functioning entity - a minute factory of life. It contains many complicated structures and systems and is constantly working. Oxygen and nutrients such as sugar are absorbed and then fed into a highly complex piece of cell machinery in order to produce energy, and unwanted waste such as carbon dioxide is expelled and carried away by the blood stream (also containing cells of course). It is only through the healthy working of all the cells in the body - and the dynamic interactions between them - that an entire creature can function efficiently.

The second and perhaps most fantastic aspect of these hard-working units is their diversification. The basic cell structure has been adapted with such imagination that it can perform a vast array of roles. For example, skin cells give us a protective, waterproof covering all over our bodies; cells in the ear conduct sound to the brain; and white blood cells travel in the bloodstream fighting off infection and disease.

In plants, cells in the leaves are adapted to capture light energy and to reduce water loss, and cells in the flowers may contain bright pigments and produce scents to attract insects. Without this differentiation, large and complicated multicellular animals and plants could never have come into existence. Life on Earth would have been confined to the relatively simple blue-green algae and bacteria.

If we focus our attention on a single facet of the body - the eye for example - we can begin to understand how each cell has its own specialised role, but it is only through marvellous team work and adaptation to duty that we get results.

Cells of the coloured iris at the front of the eye contract in response to external light conditions, restricting the amount of light entering the pupil on a sunny day for instance; specialised muscle cells finely adjust the lens inside the eye, helping to focus the image; retinal cells inside the back of the eyeball capture the different wavelengths of light that make up the image and convert them to electrical impulses; cells of the optic nerve transmit these impulses in a special code to the brain; and, finally, cells in the brain analyse the messages, forming a picture that we can understand. And all this in just a fraction of a second!

Another intricate organ which relies on specialisation and team work is the ear. Here, the cells have formed unique, individual structures that each play a vital role in the whole design. A sound wave echoing through the air is first captured by the external ear, which channels it inside the ear canal. The wave travels down in a split second and thumps onto the ear drum - causing knock-on reverberations along three specially designed bones of the middle ear. This series of bones amplifies the sound whilst conducting it to the inner ear, called the cochlea. This is a very unusual structure, consisting of a narrow cylindrical tube that is tightly coiled at one end with a bizarre head of three semi-circles at the other. And along its length there lies a membrane. This is no ordinary membrane however, for on it there sit approximately 30,000 nerve cells, each one a very delicate, precision-built electrical conductor, with an exclusive fast route to the hearing centre of the brain. Unlike ordinary skin cells, for instance, these hearing cells cannot be repaired or replaced - one too many loud rock concerts and your hearing is damaged for ever.

However specialised or well adapted the organs of the body, they still have to refer to one supreme authority for the really complex operations - the brain. Although the brain is often compared with the most powerful of computers, this analogy does it no justice

Miracles in Action

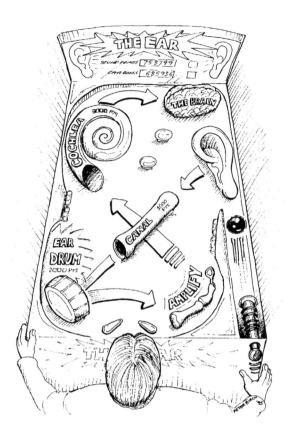

whatsoever. For the tiniest fragment of our brain is more intricately wired-up, and capable of feats far more astounding than the most hi-tech of microchips. The brain itself is a mass of nerve cells and it is these incredible power units that really operate your body, all 10,000,000,000,000 of them. Their labyrinthine, fabulously complicated connections form unfathomable electrical highways - giving you a vision, a sound, or a thought perhaps. Each one of these nerve cells has a main cell body - as most other cells do - but, in addition, it has tens of thousands of long spidery thread-like branches. These interweave with hundreds of thousands of branches from other nerve cells to form a complex embroidery in the brain, capable of spontaneous communication.

The Dictator of Life

We've looked in the mirror and seen ourselves - body parts adapted to a variety of challenging roles and all intelligently controlled by the brain. But let's just go back to our basic cell, the constituents and ultimate design of which can be found in plants and animals worldwide, from leaves and flowers to eyes and brains.

What do you picture in your mind's eye when you think of a cell? Dare I say it, but you've probably conjured up the image of an amorphous, spherical, jelly blob - a rather lame contender for the position of *Chief Building Block of All Life on Earth*. Of course the cell is not just a simple fluid-filled sac which meekly sits in the body minding its own business. We know that it is bustling with a hubbub of activity. It contains busy biological machinery which constructs proteins, manufactures energy and responds to incoming messages from distant as well as neighbouring cells.

But the most interesting player in this industrious hive of activity rests quietly amongst it all - with the all-powerful serenity that only truly great leaders possess. It is the *nucleus* - an egocentric, self-serving government which monitors and controls all of the cell's functions. Encapsulated in the nucleus, there exists a force of such monumental importance that it has been involved in the creation and development of every living species that has *ever* existed on this Earth. And, in another one of those great inadequacies of language, this awe-inspiring omnipotent material has been reduced to the household acronym, DNA (short for deoxyribonucleic acid).

The DNA molecule wants nothing less than immortality, and to this end it is perfectly designed. Unlike other molecules - such as the proteins and fats of this world - a length of DNA has the peculiar ability to replicate itself. An exact copy can be made using the original DNA as a template. This talent enables life as we know it to exist. For when a replica set of DNA has been made, a new cell for it to rule over can be created and an identical second kingdom is formed. The capacity for growth is thus given to all living creatures - a new-born infant can grow into an adult and an acorn into an oak tree - and old or damaged cells can be replaced with new ones, thus preventing our bodies from degenerating within a few months.

Miracles in Action

The most far reaching consequence of this is the continuation of life - the ability of every animal, plant and bacterium to produce new generations down the ages. New life is created by the handing down of DNA from parents to offspring. The DNA in your cells is a combination of half of the DNA from each of your parents. Their DNA came from their parents and so on, *ad infinitum*, back to distant ancestors. The very same DNA patterns that give you life and form also helped to give life to men and women who lived in the last century, the Middle Ages, 2000 years ago, 100,000 years ago, and earlier. You are a unique person in your own right, but the DNA you contain is a mixture - bits and pieces taken from thousands of individuals who preceded you - back to a vanishing point in time.

Rather conveniently for biologists, DNA is a versatile substance which takes the form of long strings. These have been given the suitably off-putting name of chromosomes. It is not for this book to discuss the nature of these amazing beasts, but suffice to say that nearly every cell of a human being contains DNA in the form of 46 chromosomes. The number of chromosomes and their content varies, of course, from one animal to another - because it is the chromosomes that ultimately define and create a creature. It is unsurprising to find that for two species of animal that look very similar, the DNA is often also very similar. In fact some species are more closely related than you might imagine. Many people are astonished to find that apes actually share over 95% of identical DNA with us.

Much of the rest of this book talks about genetic material in the form of genes. A gene is simply a stretch of DNA in a chromosome which is responsible for programming a particular protein. Looked at in general terms, genes are the ultimate in design blueprints, because they define all of our physical characteristics from height to hair colour. They may also play a role in more abstract concepts such as fear and intelligence, and some research suggests that even the ability to recognise musical notes easily - perfect pitch - may be inherited.

Unfortunately, genes don't only programme all the things you want them to, but they may also go wrong and cause problems which are completely unwelcome. Familiar genetic diseases are discussed in more detail in later chapters, but other less well characterised disorders have been linked to problems in the genes. For example, it was recently discovered that boxers with a certain version of a gene were more likely to suffer brain damage than others. This could explain why some boxers seem to suffer relatively limited effects after hundreds of fights, whereas others have serious neurological problems after just a few.

The puzzles of the genes that once seemed so completely inscrutable are now being unlocked daily. But there is still one enigma that remains unsolved, and which stands to remind us that we are only *human*. Each set of 46 chromosomes in the cells of a human body is identical. This is because the entire person grows from a single cell. But, if the DNA in each cell is exactly the same, how does an eye cell become an eye cell when it has got exactly the same DNA as a skin cell? What causes a specialised nerve cell to develop in the brain and not in the toenail? How do our bone cells form a useful skeleton and not become squishy liver cells? The mysteries of embryo development are far from solved. All that we know is, as a result of some rather well organised communication between cells, *form* is created; cells align themselves in the correct way, and somehow part of the DNA in each cell is automatically switched off during development. Only the specific section that is necessary for the specialised role of the cell is left active. So only the part of DNA responsible for programming a nerve

Miracles in Action

cell, for example, is actually working.

On this note it seems appropriate to draw to a close. We can look closely at *life* and appreciate its enormity. But we must be aware that despite our advanced biological knowledge, *life* still can only be explained in terms of DNA up to a point: if the entire complement of DNA was removed from a cell and used to create a new body in the future, would it really grow into the same person - a being with an identical intelligence as well as looks, the same quirks of personality and idiosyncrasies? This kind of research - in mankind at least - is currently enacted only in science fiction. The *cloning* of other animals, however, is *not* science fiction any longer. It is real science fact, and we examine how man has been shaping new plants and animals in the coming chapters.

Our curiosity has been aroused and there may be no turning back now - not when it comes to the manipulation of DNA and our analysis of *life*. For an insight into where this can lead us, an instruction in bringing the dead back to life, and a shocking exploration into the future, read on.

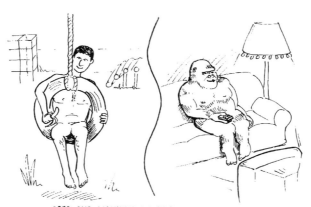

APES AND MANKIND SHARE 95% OF THEIR DNA

Attack of the Genetically Engineered Tomatoes

2

THIS IS YOUR LIFE

'The rich, thick scent of sickly sweet incense hangs heavily in the dry air, slowly curling and expanding to fill the four walls. Ancient scripts and curious carved containers adorn the high shelves. And an aged man quietly chants magical incantations and whispers mysterious prayers unto powerful gods, as he goes carefully about his business.'

The incense has a difficult job, however. Disguising the putrid festering of an eviscerated corpse is no easy feat, especially when the internal organs of the aforementioned body are lying within easy reach. But for the embalmers of Ancient Egypt, the mummification of a king was an important task and the wretched odour had to be endured.

In order to give the body immortality and to aid safe passage to the afterlife, a strict 70-day procedure was adhered to. Internal organs were pulled out through a neat incision in the body and the brain was sucked out through the nostrils. The remaining flesh was ready-salted and dried before being anointed with top-of-the-range lotions and potions, and then finally clothed in an elaborate designer suit of complex bandaging.

Three thousand years later, a curious scientist gently removes some of the carefully preserved tissue. And in a procedure which bears some disconcerting resemblances to that of the embalmer - in terms of cryptic jargon and mystical procedures - he extracts a few remaining fragments of DNA.

There is a peculiar fascination with the study of ancient DNA that has been removed from humans and other animals that have long since perished. It is the substance that gives life to all creatures, so the discovery of even a tiny segment, in an otherwise very dead corpse, is exciting. The deceased can communicate to us personal information about their lives in ways they could never have contemplated. The essence of their life is re-awakened in an afterlife undreamt of.

As well as helping us to unlock secrets of the dead, DNA from individuals who are still very much alive can be studied. Every individual has his or her own unique set of DNA - a biological type of 'fingerprint'. There is something a little unsettling about having your entire body defined in a microscopic sample of cells, but this biological reductionism does have its uses. A drop of blood or a blob of saliva can be as incriminating as a photograph, placing a villain at the scene of a crime for instance. It's not exactly a sharp, colour picture that our DNA provides, though, but more of a fuzzy, grey and white negative. The only way in which this abstract DNA image can mean anything is when it is compared to other DNA images; the only way you can be recognised from this 'photograph' is if you have provided a specimen with which it can be compared.

Scientists are still busy sorting through the hundreds of thousands of genes that make up human DNA, and they are still searching to find out what every one of them does. It may be that in the future they *will* be able to produce a colour photograph from any sample of DNA, or even a hologram! But this seems a long way off. The following chapter takes a rollercoaster ride through this brave new 'biotech' world, taking in just a few of the ways that society is deciphering and using our DNA for its own purposes.

An Ancient Pandora's Box

In 1991 a German couple happened to take a detour off the beaten track whilst walking in the Ötzal Alps between Austria and Italy. Along their route - previously untrodden by human hiking boot - they stumbled upon the brown, leathery body of a man trapped in the ice of a glacier. Little did they know, but this was not an unfortunate mortal who had lost his way during a fateful 20th Century mountain trek. This was the body of a 5,000-year-old Neolithic farmer.

The remarkable man was aptly named Ötzi, after the place in which he had lain for so many years. The remains of his body and all his

This is Your Life

'OTZI' WAS ACTUALLY A MEMBER OF THE
NORTHERN ALPS ORIENTEERING GROUP.

worldly possessions had been frozen and naturally preserved in the ancient ice - an unwitting time capsule of life. In the years following his discovery, teams of scientists and various experts pored over these remains, sampling and analysing just about everything they could, fathoming out what he'd had for dinner that last fateful night, and puzzling over the strange designs of tattoos that appeared all over his body. But possibly the most personal examination was that of his DNA.

From a small sample of body tissue, a few cells were made to relinquish the DNA that they contained, and fragments of this ancient biological code were then matched against comparable sequences taken from men living around the world. By analysing the amount of variation among the samples, the relationship of Ötzi to his fellow 20th century brethren could be determined. It transpired that the greatest difference was with Africans, the race furthest from central Europe, and the closest DNA match was made with European men, particularly those individuals living in the very same region of the Alps where Ötzi's body was discovered.

This may not seem surprising. Even with increased travel and contact between the diverse nationalities of the World, there are still notable differences between peoples of distinct regions and especially continents. But there is a deeper, more *human* agenda here. This analysis has brought close to us a man who lived over 50 centuries ago, in a world that was infinitely removed from our own. He can be seen to be connected with a living society of today. Perhaps he was even a distant ancestor of one of those scientists who eagerly sought to uncover his mysteries. In fact, in a recent discovery in the south of England, this really was the case. In 1997, genetic tests were performed on a 9,000-year-old body that had been found in the Cheddar Caves early this century. By comparing the DNA of 'Cheddar Man' to individuals living in the area, the researchers found that he has a direct descendant alive and well today - a local history teacher actually working in Cheddar!

But dead bodies don't only give us an interesting insight into themselves alone. They can also provide us with other useful information about the life and times in which they survived and ultimately perished. This is illustrated by the case of a 1,000-year-old mummy discovered in southern Peru. The body was that of a woman who had been desiccated by the arid climate after her demise - another type of natural preservation. As with a number of other mummies found in the same area, the woman appeared to have suffered from a disease affecting the lungs and chest cavity, the type of unpleasant damage consistent with tuberculosis (or TB). Nowadays vaccination against TB is given to children in the UK at an early age and although there does seem to be a worrying increase in TB worldwide, the fear associated with this distressing and drawn-out illness seems to have been forgotten. Even today, however, those afflicted must endure a harsh course of antibiotics for up to four months to ensure a complete cure.

The microscopic cause of TB is a bacterium known as *Mycobacterium tuberculosis*. In common with all other creatures, bacteria contain DNA, and scientists were able to isolate DNA samples from what appeared to be regions of TB damage in the Peruvian mummy. These were then tested against DNA samples from a living colony of *Mycobacterium tuberculosis* and found to be identical, indicating that the woman had indeed been affected by the disease.

Is this really such a startling insight? Well it certainly is for historians, epidemiologists and anthropologists. It had previously been assumed that TB, along with a whole host of other distasteful disorders, had been carried to the New World by marauding European settlers. The evidence of TB in a 1,000-year-old Peruvian mummy proves that the disease was a problem before any outside interference. The interesting question then, is ' Where *did* it come from?', to which there is no definitive answer.

If the idea that we can extract and study DNA from ancient peoples is exciting, then how much more amazing is the thought of actually resurrecting this DNA, recreating the life that remains stored within it? This concept was taken quite literally, and dramatically enacted in a film that seemed to catch everybody's imagination - *Jurassic Park*.

Based on a book by Michael Crichton, the storyline began with a hapless mosquito that had feasted on the blood of a range of animals, including dinosaurs, millions of years ago. Unaware of its future significance, the insect was trapped in a sticky plant resin which then hardened to form amber, providing a lasting tomb until the present day. The tale continued with the extraction of blood from inside the mosquito. This was the blood from its last meal on a number of dinosaurs. Dinosaur DNA was then taken from the mosquito's blood cells, amplified and eventually grown in an egg to produce new dinosaurs.

Well, how realistic is this exactly? As with all good science fiction, the fantasy *is* believable up to a point. There are many examples of insects - some millions of years old - that have been preserved in amber. In fact, ancient DNA has even been extracted from insects trapped in amber, including a 25-million-year-old bee and a 120-million-year-old weevil. The largest leap of imagination required in the story is the creation of a fully-formed dinosaur from a few strands of DNA.

It seems ridiculous now, but so many amazing DNA tricks of fiction have recently been turned to science fact that perhaps it isn't so far fetched after all. Up until early 1997 for example, it was considered impossible to produce a new, living creature using DNA extracted from living, adult animal cells. This *has* now been successfully achieved in sheep (and is discussed in a later chapter). The problem with extinct animals is that their DNA has been subject to the ravages of time and is not in a viable state to create new life. But what if it could be combined with living DNA from similar creatures alive today, to bring it back from the dead?

In 1984, pioneering researchers at the University of California were the first to successfully recover useful DNA from an extinct species of animal. Their work focused on a horse-like creature called a quagga which had roamed southern Africa until the latter part of the last century. In common with many other unfortunate species obliterated during the reign of mankind, a few stuffed specimens collect dust in museums, and it was from one such exhibit that a sample of muscle tissue was taken for analysis. The researchers managed to extract delicate DNA fragments from the dried remains, and then compared them with DNA sequences from modern-day creatures. The quagga is a very close cousin of the zebra and the two species probably shared a common ancestor that lived about 3 to 4 million years ago.

As a result of a bizarre cryptozoological brainwave, the quagga is now the subject of a unique breeding experiment. By using selective breeding techniques, a group of zebras is being biologically manipulated so that the quagga may one day walk the plains again.

Bringing the dead back to life is one thing when it involves extinct species, but what about resurrecting the recently deceased? Suppose scientists could take the DNA from our cells after we have passed away, and then use it to re-create a clone of ourselves to be re-born and live again? Amazingly, researchers in Denmark are one step down that path. They are trying to clone cows by using DNA extracted from cells of cattle which have been dead for about half an hour. DNA is collected and added to an empty husk of an egg cell, before being placed in a surrogate mother cow. If the embryo is able to develop normally, and the cow born without any problems, the new animal will be a clone of the original, dead cow. The huge implications of cloning and genetic engineering of animals are discussed in the rest of this book. But this example does illustrate how mankind is starting to play with its new found toy - the basis of life.

A Brief Interlude

A brief note should be made on the rather important technique that has enabled DNA analysis to forge ahead. This novel process has allowed us to explore new realms such as the mysteries of ancient DNA, where only fragmented segments of DNA are available. Decay of body tissue and exposure to water, oxygen and background radiation all contribute to the irreversible damage of cells, and that includes DNA. It is only against great odds that a few fragments of DNA survive devastation over time, lying dormant in tissues such as bone or muscle.

It has been the development of one ground-breaking technique in particular that has allowed analysis of ancient DNA, for example, to take place. The polymerase chain reaction (or PCR for short) sounds like the name of a bad pop group. In fact, it is a cunning procedure which can take a small fragment of DNA and generate many thousands of identical copies - a sufficient amount for analysis. It is simple enough to perform in a test tube (just add heat and various ingredients) but prone to one crucial snag - contamination. If the slightest particle of rogue DNA were to slip into the mixture, this imposter DNA also would be repeated and amplified many times. Skin cells from hands or work benches, sweat, bacteria, a cough or sneeze, cells from the previous day's work - all could be responsible for ruining an experiment. Such is the capacity for contamination that some scientists argue that ancient DNA does not exist at all, and it is simply the result of impurities in the mixture!

HEY SON, HOW ARE YOU DOING WITH THAT DNA CHEMISTRY SET ?

DNA Fingerprinting

Controversy is familiar to many branches of DNA technology but one of the more notorious topics is undoubtedly DNA fingerprinting. This revolutionary development has provided the fighters of crime with a new hi-tech, biotech means of catching their 'man'. All it needs is a minute drop of blood, semen or saliva, a hair, or just a few skin cells. DNA is such a microscopic molecule that a DNA fingerprint only requires around 50 nanograms of blood. This is incredible when you consider that an average drop of blood is about 1000 nanograms.

A full complement of a person's DNA can be easily extracted from within the cells of a fluid or tissue sample. However, rather than analyse all of the DNA available, which would take an unfeasibly long time, only about four or five segments are usually evaluated. Hidden amongst the useful regions of our chromosomes are strange sequences which consist of many repetitions, and it is these which are studied. They can be thought of as useless pages in a book, repeated many times, but the type and number of repeats varies from one book to another - one person to another. DNA fingerprinting compares these variable segments, looking at DNA from a suspect against DNA collected from the scene of a crime. If a match is made, it is likely that you have found the culprit.

There are problems, however. The technique has its weaknesses and may not be as reliable as it is often portrayed. Everyone knows that a *finger* fingerprint is completely unique. Not even identical twins have the same prints. The trouble with a *DNA* fingerprint - at least the type presented in court - is that it is not *necessarily* unique. Although the entire length of DNA for each person is different, DNA fingerprinting only focuses on a small number of sequences, not the whole DNA code. This introduces a margin of error. Groups of people in close-knit communities or within ethnic groups may have similar DNA because they share common ancestors. This means that close matches in DNA fingerprints are bound to crop up now and again, particularly within localised areas where population migration is limited. This was even the case with Cheddar Man and a local teacher, separated by 9,000 years!

The coincidental matching of DNA fingerprints is going to become more and more of a problem, and will almost inevitably lead to miscarriages of justice if the evidence is presented in the wrong light. Imagine this nightmare scenario: you are arrested under suspicion of burglary, brought to the police station and asked to supply a small amount of saliva as part of a routine procedure. Your DNA fingerprint is made and fed into a huge database that contains DNA fingerprints from convicted felons, as well as from samples left at the scenes of unsolved crimes. By an unlucky coincidence, your DNA fingerprint matches that taken from the scene of an unsolved homicide. A few hours of community service for petty theft is suddenly converted to life imprisonment for murder.

An average jury is unlikely to be a panel of experts on genetics. The techniques involved with DNA fingerprinting, the statistical jargon, and the full implications of any DNA match, may not be explained well enough for the panel of jurors to make a properly informed judgment. If they are presented with a forensic expert in the witness box who boldly states that, in his opinion, 'there are odds of one million to one against that somebody else would share the same DNA fingerprint as that of the accused', who will argue? There is evidence, however, that the average person has his doubts about this new type of seemingly infallible prosecution, and that juries may be less enthusiastic about the evidence that you might think.

The criminal trial of 1995 to which all attention turned was that of O.J. Simpson, the famous actor and sports personality who was put on trial for the murder of his ex-wife, Nicole Simpson, and her friend Ronald Goldman. By all accounts the crime scene was a rather gruesome spectacle, and the investigating officers did not have to look very hard for a blood sample for DNA fingerprinting. Blood from the victims was reportedly found not only at the scene, however. A glove that was found on Simpson's estate was apparently stained with blood that matched Goldman's, and blood on socks in Simpson's bedroom matched Nicole's. In fact, blood from all three was reportedly found soaked into the carpet in Simpson's Ford Bronco. In opposition to the seemingly damning evidence, the defence questioned the competency of the analysing laboratories, as well as suggesting that some kind of contamination had taken place when the samples were collected by police officers.

As part of the incessant barrage of media speculation and comment, an interesting poll was conducted by the American newspaper, USA TODAY, in conjunction with CNN and Gallup, to find out what people thought about DNA analysis in general. The results revealed that the majority of people did not have much faith in this type of evidence. Less than half (41%) thought it was very reliable, 35% thought it was somewhat reliable, and 8% considered it as unreliable. The rest had no opinion. And the jury in the criminal trial had an equally sceptical view; they brought in a verdict of *Not Guilty*.

There will always be the possibility that 'scientific evidence' will be misrepresented. The previous idea of a burglar who is transformed into a murderer following a coincidental DNA match is not a million miles from the truth. Convicted prisoners *have* subsequently been

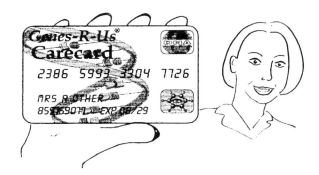

At last, you can stop worrying about health care

With the new DNA Carecard® from *Genes-R-Us*®
It contains all your health details such as ;

◇ DNA Fingerprinting
◇ Dental records
◇ Retinal Pattern
◇ Ancestral Genome History

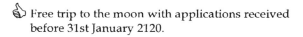 Free trip to the moon with applications received before 31st January 2120.

freed when a review of their cases has shown that DNA evidence was inappropriately presented to the jury. So this is not just an unlikely, preposterous drama. And as the DNA fingerprint database continues to grow, so too does the likelihood of accidental matches.

At the moment in the UK, only the DNA fingerprints of convicted criminals may be kept on record. Those taken from people subsequently found to be innocent have to be destroyed. How long will it be, though, before the DNA fingerprint of *every* individual will be stored on a national database? Will the DNA fingerprint become an integral part of our everyday lives? As credit card companies are battling against the ingenuity of fraudsters, what better way to protect accounts than to identify card holders by a DNA fingerprint? And would it not be a natural progression to have an identity card for all people, bearing their personal DNA code?

In a final note on DNA fingerprinting evidence, we should remember other animals and plants. In the commercial world of animal show business, DNA fingerprinting has begun to play a role - by confirming dog pedigrees for example. Just as all humans have slightly different DNA, so too do the members of all other species of animal. Although all giraffes, for example, look the same to us, their DNA varies slightly from one individual to another. They probably all look different to each other as well! The same goes for many plants, as long as they are the result of natural pollination and not cloning.

In 1993, a man from Arizona in the USA was convicted of murder with the aid of DNA evidence from seed pods. Pods from a palo verde tree, a type found growing near the scene of the crime, were found in the boot of his car. The man claimed that he had never visited the murder site. However, DNA from the pods in his car matched that from pods collected from a tree near the site, and he was convicted.

DNA evidence can also help solve crimes against animals. In 1997, for the first time in the UK, DNA tests were used in evidence against four men accused of a badger killing. Blood was found on a knife and a piece of clothing of one of the defendant's. A DNA fingerprint was made from this blood and matched to that of the dead badger. This technique can also be used against the illegal trade in endangered wildlife, matching blood collected from the scene of an animal killing to a dead creature on the opposite side of the world.

In the case of whales, the dead animals may remain on the doorstep of the country where they have been killed but are unrecognisable in their processed form of *food*. For example, some types of whale meat are still legally sold in supermarkets and restaurants in Japan, but it is difficult to tell whether the seller has obtained the meat from legitimate sources. Certain species do not have legal protection from hunters, but others such as the humpback and blue whale are protected. DNA analysis can be used to analyse samples of whale meat on sale, and check that they are from legally hunted whales and not from a protected species. The conservation group, Earthtrust, tested hundreds of samples on sale in Japan and South Korea between 1993 and 1996, and found that more than half were not from legal sources, but from protected whale species and even dolphins.

Human Genome Project

Here is another scene taken from our film of the future. The year is 2100. Peter and Jane are the proud creators of a new boy called Timothy. Timothy still hasn't begun to take on a recognisable shape yet in the test tube, but Peter and Jane are eager to know what he will look like when he grows up.

In the comfort of their own home, they tap into the computer library and download the DNA details of Timothy's chromosomes. They read with pleasure that he will have brown eyes, he will grow to a height of 6 feet tall, and he will possess above average intelligence. Finally, they slip into their virtual reality suits and are able to play with a 6-month-old baby Timothy, the creation of their dreams.

The future surely holds many surprises and developments beyond our wildest fantasies, but few could be as repellent as this. By reducing DNA - the substance which gives us our very existence - to a dictionary of information, we run the risk of losing our humanity.

The Human Genome Project is a huge international exercise which began in the late 1980s. It has as its main purpose the goal of locating and defining every gene in the set of human chromosomes - a bumper pack of information commonly known as the human genome. Literally billions of pounds are being spent so that we might one day read and understand the set of DNA instructions that are used to make a man or woman. In fact, much of the DNA code in the human genome is already freely available, accessible by the touch of a button over the internet.

Proponents of the crusade are quick to point out the benefits to be gained. The strongest argument in favour is that medical disorders could be identified more easily, and the knowledge used to find a more pinpointed, genetic cure. By a quick scan for DNA abnormalities, we could detect whether any genetic problem existed in an individual, and the symptoms treated more quickly.

This would be useful up to a point, but not all diseases have a genetic basis. And those that do are at present incurable. In some cases, even the symptoms are untreatable. Do we really want that knowledge? It is also argued that the money spent on the Human

Genome Project would be better spent targeting individual genetic disorders directly, rather than trying to map all of our genes. When you take a close look at our DNA it is a shock to find that only a small percentage is composed of what appear to be useful genes. The vast majority - possibly as much as 90% - is what is known as 'junk' DNA. This is an unfathomable mass of code that has no obvious purpose for the working of our bodies.

The elucidation of our genetic material may not only be a relative waste of time and money. It may have dire consequences. In October 1996, the World Medical Association warned that increased knowledge of the human genome could create the potential for new weapons - biological weapons - to be targeted against specific groups of people. We know that there are small differences between individuals of different nationalities. If the Human Genome Project defined these differences in detail, it could allow the development of bacteria or viruses specifically aimed at a particular race.

As well as the Human Genome Project, there is another more controversial study which appears to have less scrupulous motives and methods. The Human Genome Diversity Project has been set up to collect and preserve DNA from several hundred unique, indigenous tribes from around the world. In a remarkable biological bank, blood samples taken from these people are given immortality. Batches of blood cells are kept in temperature-controlled flasks, gently rotated and provided with sufficient nutrients to keep them alive. It is a rather macabre, unnerving thought that these colonies of blood cells can be maintained indefinitely, possibly lasting until long after the members of the tribe from which they were taken have perished.

The most extraordinary twist to the plot, however, is that the people themselves do not always know what is going on. At the same time as the researchers provide badly needed medical services for the villagers, they may take an extra blood sample from each individual, specifically for the project - although this may not be properly explained. Not only that, but it appears that certain pharmaceutical companies have also realised the advantages. The value of studying blood from these genetically distinct tribes - for research into diabetes for example - is enormous. And there are substantial profits to be made from any commercial developments.

All professionals involved in research that involves human subjects are bound, in Western nations at least, by a very strict code called Good Clinical Practice (or GCP for short). For all studies which are carried out, GCP provides a comprehensive set of guidelines which should be, and usually are, followed. One of the most important concepts is the giving of consent by a person - after the full aims and implications of the research have been explained. And in all but the most minor research, the *written informed consent* of all of those individuals taking part should be obtained.

A report by the Royal College of Physicians suggested that people offering to take part in studies should be asked a number of questions such as: 'Have you had the opportunity to discuss the study?' 'Have your questions been satisfactorily answered?' 'Have you had enough information on the study?' And perhaps the most basic - 'Do you agree to take part in this study?'

AT LAST THEY HAD FOUND IT, AN INDIGENOUS TRIBE OF PHARMACEUTICAL RESEARCHERS.

Those researchers involved with the clandestine taking of blood samples from indigenous tribes defend themselves by saying that these people could not begin to understand about DNA and genes. The researchers also claim that the taking of blood is justified because in the long term it may contribute to the greater good of mankind. A controversial proposal has recently been put forward by the Human Genome Organisation, a worldwide co-operation of scientists organised by the International Bar Association. This suggests that these special populations have a right to share in the profits of any research, if their genetic material has been used in that research. It remains to be seen whether any pharmaceutical company will act on this.

This chapter has provided just a small taste of how DNA can be taken outside of its usual resting place, from within our bodies. We can learn from it, but we can also wield it out of context and exploit it for our own ends - for good and bad. By doing so, however, we run the risk of debasing DNA - the biological basis of our lives. It is essential that society maintains a certain reverence for it and does not become blasé about its manipulation. The moral line, however, has yet to be drawn, and the following chapters describe mankind's early steps in an entirely new ball game - genetic engineering.

Attack of the Genetically Engineered Tomatoes

3

ATTACK OF THE GENETICALLY ENGINEERED TOMATOES

'I was scared, real scared. Sure, it lay there on the plate innocently enough and it was the right colour, but I had a feeling about this one. This wasn't like any other tomato I had come across before. The important thing was not to panic. I glanced sideways around the room, weighing up the situation. It was just like any other diner in downtown Chicago. The waitresses were worked off their feet, and the smell of grease and burgers was so thick you could see it.

I looked back down at my plate. This was it. I had to know. Slowly I reached down and picked up the tomato. It was a perfect specimen - firm, plump and bright red. I looked at the rest of the meal - limp and soggy lettuce, a burger made of charcoal, and the bun at least a day overdue for retirement. Everything you would expect from a place like this. But the tomato?

My mind was racing. Could there be another explanation? Surely they hadn't reached this part of town yet? But no, I had to face facts. This was no pure tomato. This one had been changed, made over, engineered to look good. But I knew it was bad, bad, bad.'

Fact or fiction? It sounds like the unlikely plot of an inept but hilarious, trashy novel. You can imagine the main character, a private investigator of course, named Dan Connors or such like. During the course of the story he would no doubt struggle heroically to overcome his terrible fear of small vegetables, a problem caused by some abominable childhood experience.

Strangely enough, however, our hero Dan would not have met such a dilemma in the real world. In 1993, Chicago became the first city in the USA to show a hint of uncertainty about a technology that was just breaking new ground - a technology that tinkers with the

Attack of the Genetically Engineered Tomatoes

fundamental genetic material of organisms, plant and animal alike. It is a breakthrough in modern science known as genetic engineering.

By law, and backed with a penalty of $500 for non-compliance, all restaurants and similar establishments in Chicago are now obliged to advertise clearly the presence of genetically engineered food. But when is a tomato not a tomato - or at least not the tomato it once was - and should we even be terribly bothered one way or the other?

Flavr Savr (pronounced 'flavour saver') is one such type of tomato. It certainly looks and tastes like a ordinary tomato, but in a number of ways it is superior to ordinary tomatoes. This particular variety has been given advantages over its brethren by a small degree of genetic engineering. By manipulation of its genes, the Flavr Savr tomato stays ripe for longer than ordinary tomatoes, which means big benefits: less produce is lost to rotting; increased harvesting and storage times provide greater flexibility for suppliers; and even the consumer gains at the end of the day with an aesthetically pleasing, relatively squidge-resistant product.

THE NEW 'IMPROVED' TOMATOES INSISTED ON DOING BALANCING DISPLAYS WHENEVER SOMEONE OPENED THE FRIDGE

Developed by a biotechnology company, called Calgene, and the Campbell Soup Company, this tomato represents an entirely new science which is going to affect our lives more closely than we may realise. From daguerreotype to hologram, or horse and cart to

Concorde, the miracles of technological progress have stemmed from the same seed - an insatiable curiosity that seems to be inherent in mankind[1]. Traditionally, this creative urge has been restricted to the nuts and bolts of non-living, inanimate objects - the computers, cars and other products of corporate technology that have transformed our lives beyond recognition. More recently, however, this burning desire to design and manipulate has reached *into* the living.

Mankind is no longer content with the existing array of plants and animals which fill every niche of the Earth. He has become impatient to improve upon that which nature has provided, in order to meet his own ends - enhancing food to taste better, generating higher crop yields to feed the expanding populations, creating pest-resistant super plants, and even making flowers bloom to order. The capability to physically engineer living things is relatively recent, because our knowledge of the intricacies of biology is also relatively new. Until one momentous day in 1953, when Francis Crick and James Watson elucidated the three dimensional structure of life, we simply did not have the means of engineering life[2]. But from that point forward there was no turning back. The biotechnological revolution that has now gained such an incredible momentum seems unstoppable, almost out of control. And yet its significance is unsurpassed, for we are now toying with the control of life itself.

This new revolution of genetic engineering must not be confused with the age-old art of breeding. For centuries, the match-making farmer has encouraged amorous encounters between his best bull and most productive cow in order to obtain better calves and greater milk yields. And in the plant world, the exchange of pollen from one flower to another is even easier - one cotton bud and a few patient weeks later, a new plant is born, exhibiting perhaps the golden hues of its mother's petals and the strong healthy green leaves of its father. In fact, if it had not been for the cultivation of 'wild' types of crops over the past several hundred years, most of the food we eat today would probably be quite unappetising and relatively inferior in quality.

---

[1] There are those who would suggest that curiosity and a healthy thirst for knowledge are not the primary driving forces. There are those who would shamelessly suggest that really man's unerring interest in technological change stems from his heartfelt desire to make life as easy as possible and if there's a chance of earning a tidy profit, then why not?

[2] As described in previous chapters, this unassuming, rather attractive molecule has been given the catchy acronym 'DNA'. Most of us have heard of it and some even know that it is the stuff genes are made of, but hardly any can spell with confidence what the letters actually stand for!

These traditional exchanges which farmers have been using for centuries are the art of cross-breeding. They *do* mix the genes of different plants or animals, but in a fundamentally *natural* manner that is sexual reproduction. It may involve some encouragement and prodding in the right direction by the hand of man, but it is still a 'natural' act between two creatures, and organisms *of the same species*.

Genetic engineering, by contrast, is a wholly 'unnatural' act. It involves the hand of man reaching into a specific cell of a plant, animal or bacterium, and locating a precise gene which may be of use. That gene can then be extracted from the huge tangled mass of DNA and put into a completely different organism. Because it is the genes in all creatures that give them their physical characteristics - whether thick, brown fur or fragrant, yellow flowers - by transferring genes we can make the recipient organism exhibit whatever features we want it to.

Not only can we transfer genes from one species of plant to another, but we can also swap genes from animals to plants and even from humans to bacteria, to create *transgenic* organisms. Imagine a sunflower 'mating' with clover, a scorpion 'mating' with a radish, or even a bacterial cell 'mating' with a human. These most definitely unnatural acts seem quite distasteful, but are no longer an impossibility.

The imagination recoils in amazement. All other scientific breakthroughs made in the past 200 years pale in comparison with the possibilities of genetic engineering. This is manipulation of nature on an impressive scale. When you change the genes of a plant or animal, you change the controlling force of that organism. Genes mean power. The entire structure and function of life is determined and regulated by genes - red roses instead of blue, two arms instead of three, Claudia Schiffer instead of the hunchback of Notredame.

Effectively, we can now manipulate life itself.

How?

It all sounds so implausible, and genetic engineering is, in fact, considerably more difficult in animals that it is in plants and bacteria. In order for it to work, the process has to be carried out very early on in the life of the organism. The gene should be inserted into the very first cell of the plant or animal recipient. This means that, as the organism develops, all of its cells will get a copy of the added gene.

Attack of the Genetically Engineered Tomatoes

Plants are easier to engineer than animals simply because reproduction is much less complicated and, in many cases, any cell can grow into a completely new plant. In animals, the whole act of sexual reproduction is more complex and rather hit and miss. The developing fetus can't be grown in a nutritious medium in a dish on a laboratory bench, and although recent developments in cloning (discussed in the next chapter) have made it possible to grow a new creature from a cell taken from the rest of the body, it is not easy. Although copies of all genes are present in most of the cells in an animal's body - the same as for plants - many of the genes are switched off in animal cells. This 'switching off' happens early in the developing fetus, so the information needed to create a new creature is not active.

Plants are easier to genetically engineer for another reason - we can insert foreign genes into their cells very easily - even by shooting them! For example, in many developing countries, the staple diet is heavily rice-orientated, but just like any other plant, the rice plant is under attack from a variety of unpleasant pests and diseases which reduce yields and hence profits. The Indica varieties of rice are responsible for approximately four fifths of rice production worldwide and scientists were particularly keen to make them resistant to a notable pest, the red stripe virus. An ingenious gun was devised which shoots virus-resistant genes into the rice plant cells. The genes are applied as a coating to special metal beads which can be fired from the gun. Once these foreign genes have been taken up by the cells into the rest of the DNA, the rice plant is able to make a special protein which helps defend itself against viral attack.

Another, more common and less expensive method used in genetic engineering recruits the services of an unlikely ally - bacteria. These are no more than annoying nuisances to most of us, entering our bodies, multiplying rapidly and causing us the discomfort of a common cold or worse. But this infective nature can be put to good use in plants. A type of bacteria that lives in the soil, known as *Agrobacterium tumefaciens*, naturally infects plants such as tomatoes and potatoes. It can get in through the roots and causes a tumour-like growth once inside the plant cells. During this infection, it manages to insert its own genes[3] into the plant's DNA. In the laboratory this bacteria can be used as a means of transferring specific genes into plants. The new, beneficial gene is first slotted into the bacteria's DNA. The bacteria are then placed in a growing medium in a dish, with the plant cells, which they then naturally infect. Once inside, the beneficial gene from within the bacteria will be transferred to the plant cells' DNA. As the plant grows, the new gene is copied into every cell of the plant, together with the rest of the plant DNA. It can then exert its effect - creating bigger, redder tomatoes, for example, or conferring resistance to a particular pest.

This method of 'infecting' the plant with foreign genes has been used for engineering numerous types of crop, and not only to increase resistance to pests and to improve yields. For those of us who have to add two heaped spoonfuls of sugar to our coffee in order to appease our taste buds, there are new developments in the pipeline which are of great interest. Researchers in the US are experimenting with the production of highly sweetened vegetables by genetic manipulation. Monellin is a protein which is hundreds of thousands of times sweeter than sugar and occurs naturally in the seeds of an African plant. Such

---

[3] Just like the cells of all plants and animals, bacterial cells also contain DNA to programme their structure and function. In addition to the genes contained in the main area of DNA (in the nucleus), many bacteria also have a peculiar circular section of DNA called a 'plasmid'. The genes contained in this DNA loop are not usually vital to the well-being of the bacterial cell, but may have other benefits, such as giving the bacterium resistance to an antibiotic. It is relatively easy to add genes to the plasmid of a bacterium, compared to the ordinary lengths of DNA found in the nucleus known as chromosomes. Hence, they are often used in genetic engineering.

Attack of the Genetically Engineered Tomatoes

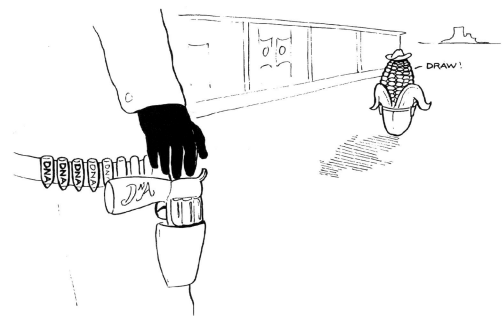

RESEARCHERS HAVE USED DNA-TIPPED BULLETS TO FIRE GENES INTO PLANTS TO HELP PROTECT THEM FROM DISEASE

a minute quantity is needed for sweetening, that the calorific value is virtually nil. The gene which is responsible for making monellin has been transferred into tomato plants using the *Agrobacterium* bacteria, to produce sweet, juicy red tomatoes. And Japanese researchers have recently put the gene into a common yeast, *Candida utilis,* which can produce the protein in huge quantities.

There are other, less obvious, products that our plants can also be engineered to provide for us. The US based company, Agracetus, has isolated genes from the bacterium, *Alicaligenes eutrophus,* which make a biodegradable type of plastic called poly-hydroxybutyrate (PHB for short). A gene gun was used to fire these genes into cotton plant embryos, and the developing plants actually manufactured granules of plastic within themselves. This kind of technology may be used to create a highly advanced form of clothing, which could perhaps retain heat more effectively than ordinary cotton fibres, or perhaps repel moisture.

Further researchers in the US are working on another use of genetic engineering - for vaccination via foodstuffs. The food poisoning bug, *Escherichia coli*, and the hepatitis B virus have both been investigated with a view to potential vaccines in the form of bananas! The genes responsible for the part of the bacteria or virus that triggers an immune response in humans are first identified, and then transferred into the food, by *Agrobacterium tumefaciens* for example. When the banana is eaten, an immune response to the bacteria or virus is stimulated, thus vaccinating the person against the disease. This type of immunisation could be much less expensive than the traditional vaccination programmes and could be extended to a range of diseases.

In Canada, researchers have been investigating potatoes genetically engineered to help prevent diabetes. Type 1 diabetes is an auto-immune disorder, which means that the body's natural immune system starts attacking the body itself. A protein in the pancreas called 'GAD' is particularly affected; GAD is present on specialised cells that produce insulin - a substance needed to regulate sugar levels - and when it is attacked, insulin is no longer produced properly. The gene that produces the GAD protein was inserted into potato plants, so that high levels of GAD were produced in the potatoes themselves. When fed to a strain of mice that were susceptible to type 1 diabetes, the disease was apparently prevented from developing; only 2 out of 12 mice developed diabetes compared with 8 out of 10 control mice which were not given the potato. It is thought that this approach may also help in other auto-immune disorders like rheumatoid arthritis; by feeding the body large quantities of the susceptible protein, it may help build up a tolerance to the immune system's attack.

It is not only the finished produce that can be affected. Genetic engineering can be used to affect production indirectly. Australian researchers have introduced a gene from sunflowers into clover plants. The sunflower gene increased the amount of sulfur-rich protein made by the clover by more than 100 times. By feeding sheep with this clover, the amount of wool the animals produced also increased!

Another example of this indirect action has been used in cheese production, and the resulting produce is already in the shops. Rennet is a substance extracted from the stomachs of calves, from which an enzyme known as chymosin (formerly rennin) is extracted. This is the primary milk-clotting enzyme used to make cheese. By genetic engineering, the gene which is responsible for the manufacture of chymosin by cells, has been transferred to bacteria. The bacteria are then used to manufacture genetically engineered chymosin. Vegetarian cheeses are now manufactured using this source of the enzyme, rather than that from the calf's stomach.

The example of cheese production illustrates the important potential of bacteria. Rather than carrying a gene into a plant, a genetically modified bacterial cell may itself produce useful substances and not only in terms of food production. Proteins which are needed in large quantities for medicines can be made in bulk by genetically engineered bacteria. Diabetes mellitus affects approximately 1 in 70 people in the UK, and many diabetics have to administer to themselves regular insulin injections in order to control the condition. In the past, this insulin had to be extracted from the pancreas of a pig or cow. Nowadays, copies of the human insulin gene can be inserted into bacteria or yeast cells. These micro-organisms are then used as tiny manufacturing facilities, producing human insulin for us. Because micro-organisms reproduce easily and very rapidly, a huge broth of miniature insulin-producing machines can be generated from just a few cells.

The Other Side

On the surface it would appear that we can expect all kinds of new weird and wonderful developments - all for the good of humanity. Our crops will be more resistant to disease and will produce higher yields. Our food will taste better, look more appetising and stay riper for longer. And we can turn microscopic bugs into human medicine factories. Even if you don't agree with the small matter of the manipulation of life itself, surely the altruistic cause that it serves is highly commendable? Can we not justify the means by the end result?

It all depends upon how you view the world. Are you an eternal optimist who believes that politicians are a noble, but simply misunderstood breed; that multinational pharmaceutical companies have the benevolent aim of developing drugs to help make people better; and that Darth Vader wasn't such a bad chap after all? Or are you a die-hard cynic who knows that everyone is in it for themselves; that multinational conglomerates are taking over the world; and that Luke Skywalker should have been killed in the first 30 minutes of *Star Wars*?

Unfortunately, judging by recent evidence over the last few years, the cynics seem to have an increasingly strong case as far as the genetic engineering boom is concerned.

Let us consider a crop that affects us all, not in terms of our stomachs, but rather our wardrobes - cotton. Cotton is big business - not only for companies eager to fulfil an insatiable demand for garish summer dresses, but also for those companies that help keep the crops pest-free. Cotton is also bug business. It is a crop that is notoriously dependent on pesticides, being prone to attack from several unpleasant types from the insect underworld, such as the pink bollworm caterpillar - a pernicious beast with a voracious appetite. The boll is the seed pod of the cotton plant, and is an attractive nesting place for the female moth. Once hatched, the larvae innocently feed on the juicy boll, burying through it and incidentally damaging the cotton fibres. Hence, the bollworm larvae are not the most welcome of guests.

In an attempt to keep the nations of the world clothed, approximately 500 million dollars is spent by cotton farmers in the US on around 6,000 tons of pesticides. The obvious genetic engineering solution would seem to be to engineer cotton plants to help them resist the pests. However, this strategy would, of course, reduce the need for the cocktail of insecticides and herbicides that is currently used. Instead, an alternative approach has been encouraged and funded by notable manufacturers of these chemicals. Instead of making plants resistant to the damaging pests and weeds, researchers have made the plants more resistant to the pesticides and herbicides that combat these problems. In this peculiar environmental twist, greater quantities of chemicals and even more toxic alchemical marvels can now be applied to crops. For example, the chemical bromoxynil is usually toxic to ordinary cotton plants as well as to the weeds which frustrate the cotton plants' growth. The biotechnological company Calgene (remember the tomatoes) isolated a bacterial gene which provides resistance to bromoxynil and introduced it into the cotton plants. These plants are now resistant to the chemical, thus allowing the spraying of crops to prevent weed growth - but with no damage to the cotton plants.

The research was apparently sponsored with the aid of the pharmaceutical company that manufactured bromoxynil. And cotton is not the only crop to have been genetically engineered to tolerate bromoxynil; the gene has also been inserted into tobacco plants, with research again sponsored by the chemical's manufacturers.

A more recent debate has concerned maize. A genetically

engineered variety has been developed with the aid of another multinational pharmaceutical giant with a strong interest - this time in a herbicide. The designer maize plant contains a gene for resistance to a herbicide called glufosinate, as well as a pesticide resistant gene to protect it against the European corn borer - a beetle whose larvae destroy around 10% of Europe's maize crop. The focus of debate over this crop has not been the genes inserted to confer resistance, but the marker gene which has also been added, alongside the resistance genes.

It is usual for manufacturers to attach a type of marker gene when they are adding beneficial genes to plants, in order to help make sure the genetic engineering has worked properly. The most convenient type of marker is a gene which gives the engineered plants resistance to a particular type of antibiotic. In order to check the genetic engineering has worked, i.e. that the new gene has been incorporated into the recipient plant, the new plants can be grown in a medium containing the antibiotic. If the new gene (and its antibiotic marker gene which is attached) has been successfully incorporated into the DNA, the plant will survive.

In the case of the genetically engineered maize, the marker gene gives resistance to ampicillin, a type of antibiotic related to penicillin. Concern was expressed that when animals ate the maize, this gene could be transferred to bacteria that live in the animals' guts, and that this could subsequently help spread bacterial resistance in humans. Initially in the UK, the Advisory Committee on Novel Foods and Processes declared that the maize was unacceptable due to the ampicillin-resistant gene. They suggested that there was a possibility of a transfer to bacteria in animals fed the unprocessed genetically engineered maize, and that it may compromise the use of veterinary medicine. However, the European Commission overruled them, and under extreme economic pressure, and for the sake of preserving trade relationships, it has now been allowed in.

A similar concern had surrounded Flavr Savr, the genetically engineered tomatoes, which also contain marker genes, that give resistance to the antibiotics kanamycin and neomycin. However, the UK government committee which reviewed the product ruled that the presence of these genes would not compromise the clinical and veterinary use of the antibiotics. This may seem surprising. The spectre of increasingly resistant strains of bacteria appears to be looming larger, all over the world. For instance, in Japan, a type of

Staphylococcus aureus bacteria has now been discovered which is resistant to vancomycin, an antibiotic which had been considered the 'last resort' option. Similarly, in the UK, *Enterococcus* bacteria that have become resistant to vancomycin have been reported in many hospitals. Even diseases which are thought of as extremely rare may once again become a menace due to the transfer of antibiotic resistant genes. For example, in September 1997, French researchers reported the isolation of a highly resistant strain of the bacteria which causes plague, *Yersinia pestis*. The bacterium was carried by a boy from Madagascar, and was resistant to all the antibiotics normally used to treat it, and also all the drugs that had been considered effective second-line therapies. The origin of the strain was unknown, but of great concern was that the multidrug resistance was easily able to transfer to other strains.

More and more research is supporting the belief that genes for antibiotic resistance are easily passed between different types of bacteria via their plasmids. It is unsurprising that concern is mounting over the potential for problems when genetic engineers insert antibiotic resistant genes into their products.

And it may not only be the antibiotic resistant marker genes that could be causing problems in the long run. In January 1997, researchers in Germany made a claim that seems to contradict what has been a commonly held belief - that ingested DNA is harmless because it is broken down in our stomachs. The researchers found that DNA fed to mice not only survived the digestive processes, but was also incorporated into cells in the liver, spleen, intestines and white blood cells. The significance of these findings is unknown. It may be that the foreign DNA would not survive for long, and that there would be no effect on the cells of the animal. But the crucial point is that nobody is sure.

From the consumer's point of view, the safety of our food is paramount. But who decides whether genetically modified food is fit for human consumption, and who governs the extent to which the public is informed about new developments? Have there been sufficient resources and time spent to prove that genetically engineered food is safe, and is anyone really looking at the possible long term effects? We must be able to trust not only the companies that develop these products, but also the governmental bodies that make decisions over safety, and ultimately what we see on supermarket shelves. In the UK there are a number of regulations that keep tabs on genetic modification in order to safeguard human health in a laboratory setting and also the environment. In addition, the Government receives recommendations from committees such as the Advisory Committee on Genetic Modification (ACGM) and the Advisory Committee on Releases to the Environment (ACRE). For matters involving new foods and processes, the Advisory Committee on Novel Foods and Processes (ACNFP) advises the Ministries of Health and Agriculture.

The first genetically modified food that was assessed, in 1989, was a type of yeast used in bread making. *Saccharomyces cerevisiae* contained new genes which made the dough ferment at a faster rate. The ACNFP decided that the new yeast was not a threat to human health and safety, and it is now used in bread making today.

Since that first approval of yeast was granted, a small number of other genetically engineered foods, ranging from cheese to salmon, have gone on sale. There are, however, no central Government figures which list the types of genetically modified foods being sold in our shops. And most importantly, at present there is no legal requirement for these types of foods to be labelled by manufacturers. The trend seems to be that companies should label food that actually contains artificially added genes, but that secondary products, such as soya bean oil, tomato puree, or meat of animals that eat genetically engineered foodstuffs, do not have to be labelled.

Although we don't all closely examine the ingredients of each item in our shopping trolleys for artificial flavourings or colourings for example, the evidence suggests that most people would like genetically modified food and its resulting products to be labelled in some way. The consumer would like to make a fully informed choice. From the manufacturers' point of view, the issue of labelling is an important one. If one type of cheese for example was labelled as NOT genetically engineered, it might be argued that there is a subtle implication that it is better or even safer than the genetically modified varieties. In addition, people may be unduly cautious about eating genetically engineered products due to a lack of information about the entire issue.

There is evidence that the public feels rather disillusioned about the biotechnology industry in general. In a recent survey published by the Centre for Technology Strategy at the Open University, 75% of people questioned felt that biotechnology companies keep quiet about what they're doing, 65% believe that they put profit ahead of morals and 72% felt that genetically manipulated products might be toxic!

There are a number of groups of people for whom genetically modified food is not merely disconcerting, but it may actually be in violation of their religious beliefs or other dietary restrictions. The consumption of pork is forbidden by Islam and Judaism, and beef by Sikhism for instance. With genetic engineering, there is the possibility of a transfer of animal genes into vegetables or other types of animal meat. These types of genetic exchanges have been termed 'ethically sensitive'. In September 1992, on the recommendation of the ACNFP, another learned body, entitled the Ethical Committee on Genetic Modification of Food, was assembled in order to consider the moral and ethical issues of genetically altered food. After a year of deliberation and consultation with various organisations, the Committee issued a comprehensive report of the wide range of opinions held by the public. Most importantly, the report recommended that all foods that contain ethically sensitive genes should be labelled as such, but ethically sensitive genes should not be used if an alternative is available.

Another suggestion of the Committee, which is more startling, proposed that there are no compelling ethical objections to the use of organisms containing copies of human genes as food. Is this cannibalism? A helping of human liver washed down with a nice Chianti would generally be considered cannibalism. But on the other hand, a few microscopic skin cells, inadvertently shed on a perfectly normal plateful of food for example, go unnoticed - even though they may be from a number of sources - the chef, the waiter, the maître-d'hotel, your romantic partner leaning intently forward over the dinner table to stare earnestly into your eyes... Each day a person sheds around 50 million skin cells, each of which contain their owner's genes, but we tend not to worry too much about what we can't see. Out of sight, out of mind. Perhaps it is the knowledge that genetically engineered food may actually contain copies of human genes that disturbs some people.

Environmental Implications

With so many complex issues and very limited public information, it is perhaps not surprising that we feel bewildered by genetic engineering. In the biotechnological survey published by the Open University, many of those questioned expressed a feeling of helplessness. There was a general impression that industry is pushing forward the boundaries of technology and there is very little that the public can do. Despite the consumer-friendly image portrayed by big businesses, it could take just a single consumer-unfriendly mistake to stimulate the underlying suspicion of their motives, and a knee-jerk reaction against all of this technology.

In the past few years a further dimension has been added to the question of genetically engineered food and other organisms - and it is one that the public may have even less control over. This is the release of genetically engineered plants, bacteria and viruses into the environment. Although the addition of beneficial genes to our crops may initially be an agricultural bonus, there is the possibility of long term unforeseen consequences.

Flora and fauna do not survive in isolation. There is a constant exchange and interaction between organisms within the natural ecosystem. What happens when a genetically altered plant is let loose in the environment? Once it begins to reproduce and spread, the artificially-inserted gene may become incorporated into other, wild plants. Seeds and pollen can easily be carried for miles by the wind, insects or birds, thus sowing the new gene in areas far from the original site. Over subsequent generations, the genes and characteristics of non-agricultural varieties of plants may be changed drastically, conceivably affecting other plants and even animals.

This potential has been demonstrated to a certain extent already. Researchers at the University of California examined the transfer of a particular gene between radishes. Pollen from a crop of cultivated radishes was found to have carried to - and fertilised - wild radish plants. When seeds were collected from the wild variety, it was found that many contained the foreign genes from the cultivated radish plants. Even plants as far as a kilometre away were found to be affected. A similar experiment conducted by a joint French, Belgian and British team has also shown that insects may be adversely affected by genetically engineered plants. A strain of rapeseed was engineered to produce proteins that protected the plants from beetles. The beetles that feed on the rapeseed are a natural pest, but were effectively killed by the protein that the plants produced, manufactured by the newly inserted gene. However, the noxious protein was also present in the pollen and nectar. These were picked up by bees visiting the plants and transferred back to the hive and the honey. The researchers found that the protein had a fatal effect, killing bees exposed to this protein months earlier than normal. Similarly, researchers in the UK found that aphids which had fed on genetically engineered potatoes passed on adverse effects to ladybirds. The potatoes were engineered to carry a protein from snowdrops to kill off the aphids. Subsequently, female ladybirds that were allowed to feast on these aphids were found to live half as long as females fed on other aphids, and their reproductive capacity was also reduced.

If the potential for adverse effects on insects and other plants is not worrying enough, then the possibilities for micro-organisms surely are. The transfer of genes between different types of bacteria, particularly those that live in the stomachs of animals, is well known. But a Canadian researcher has recently demonstrated the transfer of genes between 2 different viruses living in a plant. One of the viruses which was used to infect the plant lacked the ability to move among cells, but the other type could. Within 10 days the virus that should have been immobile was able to move from cell to cell, indicating that it had acquired the necessary gene from the other virus. If this kind of transfer is as easy as it seems, we must be extremely careful when viruses are genetically engineered to carry certain kinds of genes. We must be aware of the potential consequences if a gene is transferred to types of viruses which we didn't intend to be affected; they may be made more virulent perhaps, and we may end up with a worse problem than we started with.

In fact, there are fears that genetically modified crops may eventually cause more problems than they solve. For example, if a herbicide resistant gene is transferred from crops into weeds, it will help the weed survive! The herbicide will then no longer be able to keep weeds at bay in crop fields. If pesticide resistant genes are transferred from crops into many other plants, this could also pose problems. Pests may become resistant to the pesticide more readily and will be able to counter-attack the resistance. Ultimately, this would mean that the genetically modified crops will be susceptible to the new improved pest. In another worse-case scenario, the modified plant may simply wipe out the wild type. With its adapted genes, the new variety may have advantages which allow it to compete with the wild plants, possibly even driving them out of existence.

One thing is for certain, once a genetically engineered organism has been released, there is no going back. It would be extremely difficult, if not completely impossible, to retrieve the organism and all of its offspring from the wild. Seventy-seven percent of respondents in the Open University survey felt that the genetically modified organisms may take over and damage the environment, and 92% felt that the hazards might not show up for years.

At the moment, the release of any genetically modified organism, be it a plant, animal or micro-organism, has to be approved by the Government. Depending on how dangerous the organism is thought to be, any application has to follow one of three routes. For organisms that are considered hazardous such as dangerous pathogens or organisms for which there is no precedent, a lengthy questionnaire has to be completed and the application is considered by the Department of the Environment in consultation with ACRE, before making a decision approximately three months later.

For organisms that are considered a low risk, a new 'fast track' procedure has been introduced whereby researchers can ignore any questions on the application form which they think are irrelevant, and they will only have to wait about one month for a decision on whether they can go ahead and release the organism. Many of these so-called 'low risk' releases are crop plants which are considered not to pose a significant threat to the environment. The third route applies to organisms that are not in the 'low risk' group, but which are not considered dangerous. Researchers wishing to release these types of organisms will have to wait approximately 50 days for a decision.

The potential for a genetically modified organism to disrupt other creatures in the environment should and must be assessed thoroughly, in every individual case. It is difficult to understand how the experts can be certain in classifying organisms as high or low risk, when it seems that the very nature of the risk is unquantifiable. How can we be sure that *every* eventuality will be thought through? What happens if an unforeseeable interaction or effect occurs?

The entire arena of genetic engineering, its possible effects on humans and the world we live in, throws up many questions - not all of which have answers. It is probably the most dynamic, expanding field of research and it is taking place all over the world. Pharmaceutical companies are spending *billions* of dollars every year in research and development connected with biotechnology, and most of the international giants have acquired small, specialised 'biotech' companies. The danger is that this drive for success and profit will overide the rational debate. It is estimated that over 30 million acres of genetically engineered crops were planted in 1997 worldwide, more than double the amount planted in 1996. Giant agrochemical companies are lobbying the media and law regulators, apparently trying to gloss over the very real concerns that international consumer groups are expressing, in order to preserve the hundreds of billions of dollars of business that genetically engineered crops represent.

The manufacturers of genetically modifed crops argue that all sorts of benefits will be gained. For example, less herbicides and pesticides will be required overall, because the crops will be targeted with specific chemicals that they are resistant to. History and common sense tell us that the advantages will probably only be short term. There will always be a small number of weeds or pests in the environment that are resistant to any herbicide or pesticide (or antibiotic in the case of bacteria, as we shall see in the next chapter). By using specific chemicals only, we will be effectively *encouraging* these resistant strains. Over time, this genetically driven farming will select for the survival of those very weeds and pests it initially sought to destroy.

The ray of hope at the moment for a sensible braking of this runaway food revolution comes from governments. At the end of 1997, ministers in Britain, at least, temporarily blocked licences for growing genetically engineered crops such as oil seed rape. Although they are under tremendous pressure from large international conglomerates, they must remember that the public are unsure - our perception of the technology, and those who control it, tends towards distrust.

We need to be told more; we need to feel in control; and we must be guided by common sense and reason, not by money.

4

DO SHEPHERDS DREAM OF TRANSGENIC SHEEP?

'The raiding parties waited motionlessly. Cool and alert. Hundreds of bodies hidden from view and perfectly stationary, biding their time but ready to strike at any moment with military precision. A scout had returned the previous day with information. This was a perfect settlement and the hot summer's day was ideal timing. Although heavily guarded and with strong defences, the camp would be easily defeated. Prisoners would be quickly taken with the minimum of effort. They would be dragged back to base and enslaved - trapped for the rest of their natural days to serve a race far superior to their own.

The order came. The first group of attackers moved into place. It was the same tactic every time - a decoy group attaching from the side, whilst the main groups remained hidden outside the main gate. Suddenly they charged in - assaulting with a ferocity and attacking any defenders that dared to put up a fight in their quest to find the young. For it was the new blood that was to be captured, the youngsters that could be turned to join their forces. As the defenders realised that the fight was worthless, they gathered the young in their arms and turned to flee out of the settlement. But this was their fatal mistake. The ambush was ready for them - lying in wait on all sides to capture their precious young, to steal them from their arms.'

These were extraordinary scenes - a truly dreadful slave raid carried out with careful planning and direction. And these were no ordinary victors. This tribe is not listed in any history book that describes the battles of old, because these aggressors are still amongst us. And they go by the name of *Formica sanguinea*.

The Blood Red slave-making ants of Europe and North America.

The bizarre and astounding designs of nature can be found everywhere in the Animal Kingdom. They seem to pervade every environmental niche of the Planet, providing a limitless supply of fascinating material for the many and varied wildlife documentaries and books. The breadth of ingenuity required to dream up such fantastical creatures is inconceivable. Yet all have been designed somehow. Whichever theory of creation you prefer - a blind evolutionary force or an omnipotent power with a stupendous imagination and not a small sense of humour - one thing is for certain. In a global talent contest for Most Interesting Animal of the Millennium the judges would need more than a brief interview with each contestant and a posing parade in a swimsuit.

Yet again, however, we find ourselves in a position where the diversity which surrounds us is apparently not sufficient for us. In the previous chapter we looked at plants and how man is constantly seeking new ways to improve upon that which nature has provided and to manipulate it for his own ends. Now he is turning his hand to animals. Man's pre-occupation with 'adjusting' the natural balance of nature is nothing new of course. For many thousands of years, ever since his early farming days, man has been using his fellow creatures to suit his own purposes, and he still does so today. From the prime, healthy cattle that feed us, to the diverse array of dogs and cats which complete our individual domestic havens, the animals that surround us have been selectively bred and altered over hundreds of generations.

The next step forward in this conquest of nature has occurred relatively recently - within the last two decades. The importance of this new development cannot be overstated because it goes far beyond even the most advanced farmyard breeding programmes. It is perceived by scientists as the way forward - to solve the world's food problems, to provide better medicines, to help save lives, and to make vast profits. It is the ultimate in experimentation with animals - genetic engineering.

The public has always surveyed animal experimentation with a wary, but essentially permissive, eye. The existence of drug testing on animals is held within our peripheral awareness and is tolerated by most of us, excused by a general perception of the greater good for mankind. Indeed, none of the medicines in our bathroom cabinets, from paracetamol to cough mixture, would ever have been developed without a series of tests on animals. Hundreds of thousands of animals are used every year to assess the effects of high or lethal doses which are experimentally injected, force fed or applied to some part of the anatomy.

The phenomenon of genetic engineering is in a different league to the breeding programmes and laboratory experiments we are familiar with - although some people claim that it is merely an extension and combination of the two. In fact, there are two important distinctions. Firstly, it is a form of experimentation of a wholly *unnatural* kind; unlike traditional breeding programmes which are fundamentally natural, genetic engineering tampers directly with the very substance that gives us life - DNA. Secondly, it goes far beyond using animals as laboratory 'test' animals; we have begun to use artificially altered animals as manufacturing facilities for our medicines, as well as models to study human disease, and living animals may one day provide replacement organs for our own bodies.

The first indication that something remarkable was lying in wait came about a decade and a half ago, although few could have envisaged at the time just where it was going to lead. In December 1982, there appeared a picture of two white mice on the front cover of a leading scientific journal called *Nature*. An experiment on mice is not normally newsworthy - there are over one million experiments on mice in the UK every year. But this one was special. One of the mice in the picture was twice the size of its litter mate. It was one of the world's first genetically engineered animals. At an extremely early stage of its embryonic development this mouse had been given a gene from a rat. This gene was responsible for the production of growth hormone. As the mouse grew, the rat gene exerted its effect, causing the mouse to grow to a much larger, rat-like size.

Nearly 15 years later, in March 1996, an even more astounding achievement was announced. The headline of *MONSTERS OR A MIRACLE?* was splashed across the front page of the *Daily Mail* accompanied by a large colour picture of two slightly bemused-looking and rather shaggy sheep, standing side by side. Morag and Megan were Welsh mountain sheep with a difference. Well, actually, *not* with a difference. These sheep were not just very similar, they were genetically engineered to be identical *clones*. At a pioneering research centre near Edinburgh, called the Roslin Institute, a dream took one step closer to reality. It is a dream in which flocks of identical sheep and herds of identical cattle roam the countryside. In this dream, not only do the animals look the same as each other, but each animal has a fixed

SHEEP OF THE FUTURE WILL PRODUCE THE WORLD'S FIRST GENETICALLY ENGINEERED WOOLLY JUMPERS

milk protein to fat ratio, all are resistant to disease, and all are guaranteed to provide the farmer with profitably high meat, wool and milk yields.

The cloning of animals is the most recent and perhaps most extreme example of genetic engineering in animals. The starting point of the two sheep at the Roslin Institute was one ordinary embryo that had been removed from a ewe. Individual cells were extracted from this embryo and multiplied in the laboratory to give a number of identical, cloned cells. Because all of the cells were from the same embryo, all contained identical genetic information, that is, exactly the same genes.

For the next stage, the services of a number of surrogate ewes were required. Unfertilised eggs were taken from each of these females and all the genetic material was removed, leaving empty 'husks'. DNA from the cloned cells were added to these husks to make complete eggs, and these eggs were then replaced in the surrogate mothers. Each of the surrogate ewes was therefore pregnant with her own egg, but this egg carried the cloned genes from the original ewe. When the lambs were born the researchers could tell instantly that the experiment had been a success. The original mother had a white face, the surrogate mothers had black faces, but the lambs had white faces!

Both of these examples - the insertion of a rat gene into a mouse and the cloning of animals - illustrate a crucial point. Genetic engineering is not simply an extension of the breeding process or any natural mechanism. It is a completely artificial means of tampering with life. It is the very nuts and bolts of life itself that are being manipulated - the genes.

Cut and Paste

When a gene is taken from one species of animal and placed in the DNA of another, the new creature is called *transgenic*. Many experiments have already taken place using a range of different animals. Pigs, sheep, goats, baboons, fish and even chickens have been subject to the new molecular biology craze that's sweeping the world, although it is still mice which are most successfully and commonly used. In order for a transgenic animal to be produced, the foreign gene has to be inserted into the very first cell that is later to grow into the animal. It would be no use to inject genes into a fully grown animal because they would not be taken up by every cell of the body. For the first stage, the experimenter has to obtain freshly fertilized eggs from the reproductive system of an animal before they have had time to even divide once. These eggs are placed in a dish on a laboratory bench and then injected with the new gene under a microscope. The diameter of the glass needle used for injecting is just 0.75 µm across, - a thousand times smaller than an ordinary needle. The eggs are then returned to the animal and allowed to develop 'normally'.

The technique is not foolproof, however, and the success rate is not very good. In fact, only 5 to 30% of the embryos survive to reach full term. And in only 10 to 20% of the new-born animals will the implanted gene have successfully been incorporated into the rest of the animal's DNA. However, once the transfer has been accomplished, the transgenic animals can be bred with one another to produce a line of animals which will all contain the artificially introduced gene.

As with all experiments on animals, there are strict laws in the UK which regulate the work that goes on. Under the Animals (Scientific Procedures) Act, 1986, all work must be assessed to establish the extent of suffering of the animal and what benefits to mankind the research is likely to produce. Before a transgenic animal can be produced, there is a lengthy process which requires the researcher to fully justify the work that is planned, and a decision must be made by the Home Office before a licence can be granted. However, despite the reassuring promises, there are a number of issues that are not addressed by the legislation, and there are moral problems which are not in the remit of the law.

The obvious question mark hangs over the possibility of unforeseen long term consequences of adding foreign genes to an animal. Will there be any detrimental effects on the animals themselves or the products they give us? And apart from the safety issue, are we really sure that we want to start swapping genes between animals that would never normally mix in nature, or inserting copies of human genes into an animal's genetic material? Is it really ethically sound to use animals this way at all, merely for our own benefit? To answer these questions, it is necessary to take a closer look at what this technology can offer us, what exactly is involved, and what types of research are being undertaken.

Factories of the Future

The benefits to be gained from producing transgenic animals may not be immediately apparent. What possible benefit could there be from transferring genes from one animal to another? The answer is a great deal. And the potential for making some people somewhere very rich indeed is reflected in the range of flourishing biotechnological (or 'biotech') companies. The commercial prospects for the genetic manipulation of animals have been quickly realised and seized upon by many major pharmaceutical companies, which are increasingly forming collaborations with the smaller biotech companies. The current production of medicines is a costly process, requiring huge upfront investment in production, as well as running costs for large factories, lots of equipment and staff. The genetically engineered animal offers an alternative to this. It is the star performer in a new, booming manufacturing science known as 'pharming' - meaning a combination of pharmaceutical production and farming of animals.

Basically, pharming is the production of pharmaceuticals in living creatures.

And what a wonderful manufacturing facility these animals make. They are easily reproducible and pass on their manufacturing abilities to their offspring. They consume their own fuel supply. The production capacity can be altered according to demand, depending on how quickly the animals can breed. And the final product can be easily collected in large quantities from the milk of the animals, with seemingly no danger to the animals' health. Already, researchers have been successful in rearing genetically engineered rabbits to produce milk that contains calcitonin (a bone-building substances that helps combat osteoporosis), goats to produce milk containing a monoclonal antibody that could be used to fight cancer, and cows to give us milk containing human lactoferrin (a natural antibody found in human breast milk that could be put into infant formula).

DOWN ON THE PHARM

The potential benefits to human society are enormous. But how far are we prepared to go? Tracey and Spencer are sheep with a faintly human touch. Developed with the financial aid of a biotech company, Pharmaceutical Proteins Limited in Edinburgh, these animals have been engineered to carry copies of human genes. Humans naturally contain a gene which produces a protein called alpha-1-antitrypsin

(AAT). However, in about 1 in 200 people, there are problems with AAT production, leading to disorders such as emphysema, an unpleasant and potentially fatal lung disease. Copies of the human gene which produces AAT were microinjected into over 100 sheep eggs. Five of the resulting offspring were found to have successfully incorporated the human gene copy into the rest of their genetic material. These included Tracey and Spencer. As AAT is produced in the milk of the female sheep which carry the human gene, it can only be extracted from female sheep. Tracey turned out to be an excellent producer, who generated 35 g of AAT in every litre of her milk. And this could be a real money spinner in the long run. By the year 2000 it is predicted that a ton of the protein can be produced by sheep every year, worth a staggering 100 million US dollars.

Pharming is not the only use for genetically engineered animals. In a hatchery in Loch Fyne, Scotland, a series of experiments has been performed in salmon. In this case, it is the salmon themselves that are feeling the effects of the artificial gene. Salmon eggs were injected with growth hormone genes and other DNA from a fish called an ocean pout. The extra genes caused the salmon to grow at up to 10 times their normal rate, reaching maturity in about two years instead of three. This is terrific for salmon farmers - there are less costs in terms of pond maintenance, there is less time for animals to be lost to disease, and a product is brought to market in a much reduced time, thus increasing turnover and profits. There are also benefits, of course, for society. The boost to this form of 'aquaculture' could be used to prevent overfishing and, by increasing productivity without relying on traditional sea-fishing, the price of the fish may be kept relatively low. Those in favour argue that this will be especially important in the long term when demand will be high with increasing populations.

When there appears to be no suffering to the animals concerned, it could be argued that these kinds of developments in genetic engineering are justified. But what happens when the animals are affected? The ethical problem then becomes much more recognisable. Perhaps one of the most controversial developments in the field of transgenic animals is a pig named Astrid. Astrid represents an entirely new reason for producing transgenic animals. This pig is not intended to provide us with new and improved pork, or with disease-fighting proteins. Astrid has been developed as a prototype organ factory.

The problem with the human body is that bits of it are frequently going wrong, usually at the most inopportune of moments.

Improvements in surgical procedures and the discovery of immuno-suppressing drugs over the last 20 years have facilitated transplantation between humans, and we are now familiar with transplantation of livers, kidneys and even hearts, for example. There is just one major problem - the demand for replacement organs outstrips the available supply. In fact, the shortfall is literally thousands of organs. It is estimated that 6,000 people in Britain are waiting for a donor organ, and 150,000 people world-wide are in need. This led to an astounding situation in developing countries where kidneys were bought and sold, and tens of thousands of dollars changed hands for just one organ. Only in 1995, for example, was it made illegal in India to sell organs. In China, even more bizarrely, organs are said to be taken from executed prisoners. Sufficient organs to transplant all of the needy are apparently recovered from an unknown number who die from capital punishment every year.

As we now cannot obtain enough human organs, in this country at least, researchers have started to consider animals as a source. There have been sporadic attempts at transplanting animal organs into humans from around 20 years ago, but interest has increased in the past decade with improved techniques, and a new hope that it can actually succeed. Transplants across two different species are termed xenotransplants or xenografts. If this technology overcomes all the difficulties, in the future there will be an instant, limitless supply of organs ready to be transplanted to order - all the patient would need is to be measured up for size, and a ready-made organ would be available immediately.

The most likely candidate for the organ factories of the future is the humble pig, for a number of reasons. Pigs are already domesticated and easily bred, producing about a dozen piglets in each litter relatively frequently. Also, people are familiar with eating them and are less likely to be offended by their use, compared with primates such as chimpanzees, for example. Their organs are roughly the same size as our own and, in theory, there is less likelihood of disease transmission because they are relatively unrelated to humans. Although primates are more closely related to us, they are not a commercially viable option, and neither are they considered an ethically viable option.

Mankind has a tendency to assign different values to different species of animals. We believe that we can attribute different levels of intelligence to an ape and a pig, for example; we are happy to eat one but not another, and we think that their perception of the world around them - including suffering and pain - differs as well. Clearly, some species are more equal than others in the eyes of scientists involved in research, and also in the eyes of those who sit on committees designed to oversee the research. In early 1997, the UK Department of Health Advisory Group on the Ethics of Xenotransplantation published a report, *Animal tissue into humans*, which decided that it would be ethically unacceptable to use primates as a source of organs, because it would expose them to too much suffering. Pigs, on the other hand, *were* deemed to be acceptable creatures, as long as they were not allowed to suffer unduly (whatever that means).

Although Astrid, developed by Imutran, a company in Cambridgeshire, is unlikely to become an organ donor herself, she has paved the way for others like her. Injected with human DNA when she was an embryo, her organs were made to appear as human as possible, and hence are less likely to be attacked by a human immune system. By inserting copies of certain human genes into the pig's genetic material, protein markers on the pig's organs will identify them as human rather than pig organs. When transplanted into a human body they would not be recognised as 'foreign' and therefore would not be attacked by the immune system and rejected. If an ordinary organ from a pig were to be transplanted into a human body it would be rejected within a matter of minutes, even if the recipient was given strong immunosuppressing drugs.

Apart from the acute rejection, there is another problem which has to be assessed thoroughly before clinical trials with human beings can go ahead. It is potentially a hazard that could affect not only the health of those individuals receiving the transplant, but also the rest of society. It is one of infection. Although the pig is relatively distant from man, genetically speaking, there is still the risk of a virus or other infection being transmitted via the transplanted organ. Researchers in London recently found that a retrovirus of pig origin was able to infect and reproduce in human cells. This has huge implications. Retroviruses are life-long infections, and may be transferred by blood or sexual contact. They could easily, therefore, get into the population at large. If a contagious infection was inadvertently transferred by this new breakthrough in transplantation, it could unwittingly wipe out more lives in the long term than it could ever save in the short term.

Much of the work with transgenic animals is still very much at an experimental stage. But there are other forms of animal engineering which are up and running and which have been affecting our lives for a number of years. One of these involves a hormone called bovine somatotrophin, or BST[1], which has been genetically engineered and then injected into cows. BST is a natural substance, produced normally by the body of a female cow to control lactation. But researchers have now managed to artificially produce a genetically engineered version and by injecting this into cows, milk production can be increased significantly - by 20% or more.

On the face of it this is good news for farmers. More milk equals more profit. But there are a number of potential problems, particularly for the cows. It is claimed that the hormone increases the animals' susceptibility to a painful udder infection called mastitis. The incidence of the infection in cows injected with genetically engineered BST may be over 50% higher than in cows not given the hormone. There may also be a threat to human health. Mastitis in cows is treated with antibiotics. Although the evidence is inconclusive at present, it is suggested that these antibiotics can be passed on to humans who drink the milk, and it is well known that bacteria are becoming increasingly resistant to antibiotics as they receive more exposure to them. Yet again, the long term negative consequences may greatly outweigh any short term benefits to be gained. The trouble is, we just don't know for sure.

At the moment there is a ban on the use of BST in Europe but its use has been approved in the USA. Consumers in the USA are not even given the ability to distinguish between milk produced by cows that have been injected with genetically engineered BST and other milk. The American Food and Drug Administration (FDA) has effectively prevented manufacturers from stating on labels that the milk is hormone-free. It is argued by those who are against labelling that any labelling could be perceived as misleading, possibly implying that BST-free milk is somehow healthier or more safe. This is not consistent with other types of product - organic and additive-free products, for example, are freely labelled as such, possibly suggesting that they are better in some way than pesticide-treated or more artificial products. As with genetically engineered food of plant origin, a lack of labelling denies the consumer his basic right to know what he is eating, and it is unacceptable.

---

[1] As opposed to BSE (bovine spongiform encephalopathy) - a nasty disease which affects these animals and has been given the nickname 'mad cow disease'.

Rights and Wrongs

There is no doubt that genetic research is leaping ahead at speed. Many hundreds of companies worldwide are pursuing the profits of this monumental cut and paste technology. But the drafting of industry guidelines and the updating of legal regulations - conceived in bygone eras which could not have dreamt of these developments - has been slow. New concepts have to be addressed and tough decisions, which involve ethical and moral issues as well as practical ones, are being made. The danger is that these foundation-laying decisions are rushed - debated by committees behind closed doors and with the pressures of industry from all sides. This is illustrated by yet another important aspect of genetic engineering - patenting.

As with human genes, the patenting of genetically engineered animals is a tricky area. Should researchers be allowed to own the rights to an animal if they have manipulated the genes in some way? Despite vehement opposition from many groups, the door for animal patenting has already been pushed ajar. Both the USA and European Patent Offices have already granted patents on animals.

The first animal to be patented for its modified genetic wardrobe was a type of mouse called an 'onco-mouse'. It may look like any other mouse, but this doomed creature is genetically programmed to develop cancer within about 6 months of birth. The term 'onco-' comes from the word 'oncogene' meaning a cancer-causing gene. The oncomouse, created by researchers at Harvard University in the USA, contains a human oncogene. By inserting the human gene into the DNA of mice embryos, the animals were transformed into an animal model specifically created to study human cancer, and to test potential treatments for cancer. Researchers claim that it will reduce the number of other animals that have to be used in the quest to find anti-cancer drugs, because it has been specifically 'designed' to replicate the illness.

The patent for the onco-mouse was granted in the USA in 1988 relatively easily. In Europe, however, the application was initially turned down; the European Patent Office argued that patents on plants and animals were not permitted by Article 53 of the 1973 European Patent Convention, but this decision was later reversed in a controversial re-think. It was subsequently decided that the onco-mouse's usefulness to mankind, in terms of its value to cancer research, *did* outweigh the cruelty to and suffering of the animals themselves. Clearly not everyone would agree with this assessment. To many people the mere idea of tinkering with an animal's genes is repulsive, let alone genetic engineering to create pain and suffering - even if it is to help combat the pain and suffering experienced by many hundreds of thousands of individuals who have cancer. But man has claimed species superiority and put his needs above those of all other animals, not only by manipulating them genetically, but also claiming the right to their lives in the form of patents. These genetically engineered animals are considered inventions of mankind rather than living products of nature. We have assumed property rights over them, and their genes, in much the same way as we have plants or inanimate objects.

We have also begun to consider human genes in the same way. In 1997, the European Parliament backed a directive that allows researchers to patent human genes and other parts of the body, as long as they are 'isolated from the human body or otherwise produced by means of a technical process'. It seems incredible that companies can *own* biological tissues and molecules that occur *naturally* within our bodies. And they can even take samples from our tissue, and use it in their research. The public's attitude to this kind of control and ownership appears to go against that of the legislators. Out of a 1000 people questioned in a poll commissioned by the Genetics Forum in 1997, nearly 700 stated an objection to their tissue being used in genetic research, even if it was anonymised, and also to it being sold to drug companies without their consent.

The engineering and patenting of animals has only just begun. Cancer is not the only disease that can be mimicked in animals. It seems that for any disease in man that has a genetic origin, an animal model can be developed. Disorders such as cystic fibrosis, sickle-cell anaemia and Alzheimer's disease have all been artificially created in animals by genetic modification. Even problems that are not obviously genetically linked, for example, memory, awareness and aggression, have already been investigated. By breeding animals that consistently develop an inherited disorder, researchers can have a steady supply of subjects for their experiments into human disease. In what are termed 'knock-out' animals, particular genes in the cells of an animal are disabled in order to study what the effects are and to study new therapies. A company in Texas has plans to produce a half a million different strains of mice in the next 5 years, each with a single defective gene. Each of the hundreds of thousands of mice will have a defective equivalent of a certain type of human gene.

This chapter has attempted to provide a broad range of examples of the genetic engineering of animals. It has highlighted a number of important issues that we should be debating as a society. Should we really be interfering with animals in a genetic manner and, if we should, then how far is it ethically sound to go? And how far is it safe to go? We must be satisfied that we are completely justified in swapping genes between species that would otherwise never mix in the natural world. This is particularly the case when the animals we create have to undergo any degree of suffering. It is difficult to say what is right and wrong, what is acceptable and unacceptable. And does a sheep cease being a sheep when it doesn't look like a sheep any more; when it doesn't act like a sheep any more; or when more than 5% of its genes have been derived from a man? Adding one human gene to the hundreds of thousands of genes that define the animal may not have a humanising effect, but there could be long term consequences. And do we have the right to patent an animal on the basis of its DNA, bearing in mind that we are not the creators of this substance, merely the manipulators? There are so many questions, none of which has a completely satisfactory answer. Even with advances in the techniques, there are still strange consequences in genetically engineered animals that are unexplained, such as increased birth weight, genetic mutations and deformities, and even a potential for excessive ageing has been suggested.

There is no doubt that science needs to find cures for a great number of diseases, and the financial rewards for industry may provide a driving force for new developments. But many people are unhappy at the thought of genetic manipulation of animals - and the thought of this technology in the wrong hands. The apparent ease with which genetic material can be cloned and transferred from one cell to another is quite astonishing and raises all sorts of questions. Inevitably, one of the concerns in people's minds is the possibility of it happening in man.

As if their progress into cloning with Morag and Megan the sheep wasn't enough, the Roslin Institute researchers went a step further in February 1997, by completely turning science on its head again. Now the field of speculation about human cloning really is open. In their previous cloning experiments, DNA was taken from very early, embryonic cells. In their latest experiments, the DNA was taken from a fully differentiated cell of an adult animal. All the cells in an animal's body, be it a sheep or a human, quickly become specialised for different roles - a skin cell or a brain cell, for example, have completely different functions; only the section of DNA that is necessary for that particular cell is kept switched on. It had previously been thought that once a cell had become adapted to its particular role, it could not revert back to its early state, where all of the DNA is switched on. Dolly the sheep is living proof that this is a fallacy. Cells that were taken from a mammary gland of one sheep were made to revert back to a state where all the DNA was active, by placing them in a state of hibernation. When the DNA was removed and replaced in a completely new 'husk' of an egg cell, and then put in the womb of a surrogate ewe, it grew naturally. The perfectly cloned sheep, named Dolly, was born to an unsuspecting world.

A few months on and the next scene in this drama was played. This time the researchers went one step further by inserting a human gene into a sheep's DNA, and then producing a clone. This cloned sheep, named Polly after her breed (Poll Dorset), contains the human gene in every cell of her body, and can produce large quantities of the (as yet un-named) human protein in her milk.

And what about human life? In a slightly different scenario in the US, at least one researcher claims to have sustained cultures of human embryo cells grown from aborted fetuses. By letting these early, undifferentiated cells divide and multiply in a nutrient-rich environment, they can survive almost indefinitely, and they retain the ability to develop into any type of cell or tissue. Theoretically, they could be used to form grafts perhaps, or to replace any sort of tissue in the body. They could even be inserted into another fetus, effectively *mixing* the genes of at least two individuals. And this has, in fact, already been tried out with mice - and using *genetically engineered* cells that were cultured.

The implications of all of this research are enormous, the most obvious being the risk of human cloning. The technology could, in theory at least, enable human beings to make clones of themselves. In the UK, this would be prohibited by the Human Fertilisation and Embryology Act (1990), but the legislation in many other countries needs to be just as prohibitive. As a swift reaction to the developments made at the Roslin Institute in Scotland, both the European Parliament and US authorities did move to produce directives forbidding human cloning. But they have refrained from stopping cloning of animals; the European Commission's advisory panel on bioethics decided that animal cloning was to be permitted as long as 'unjustified' suffering was avoided - this being despite the fact that the European Union officially recognised animals as 'sentient beings' earlier in the year.

Perhaps the arguments for and against cloning are effectively redundant. Even with legal safeguards in place, now that the technology exists there will always be the potential for a less ethically minded scientist, institute or political regime to carry out unscrupulous work. We cannot know for certain that secret research into this *isn't* being carried out somewhere in the world. The most we can do is to *understand* the subject, to be *aware* of the possibilities, and to *know* where we want to draw the line.

5

IGNORANCE IS BLISS?

We all know it's out there - lurking in the shadows - but somehow we manage to put it out of our minds. As we go about our daily business, moaning about work, struggling to get up on Monday morning and fuming about being stuck in the longest checkout queue at the supermarket, the possibility of some really terrible, life-altering illness doesn't usually cross our minds.

There is, of course, an unpleasantly large array of diseases on offer throughout the world, not to mention a number of other medical problems and injuries that are ready to afflict our surprisingly vulnerable bodies. From malaria to AIDS, or a car crash to an encounter with a mad axe-wielding murderer, we all accept that they're out there and that they happen to *somebody*. The reason we can gloss over them every day without worry is because they can all be blamed on something - and something we believe we can control. If we take our malaria pills regularly, if we don't sleep around without protection, if we drive carefully, or if we stay out of dark alleyways late at night - then we can surely feel secure and safe from these evils?

Now imagine there was something lurking within your body that you didn't have control over - a time bomb ticking away a countdown to an early death or, at the very least, disability. Even worse - suppose that you knew the exact moment when this time bomb was going to go off.

What would you do with your life if you knew you had 10 years to live?

One year?
A few months?
Or even a single day?

Would you choose to try and live as normal a life as possible, maybe not even telling your friends and family, or would you set out to accomplish all those dreams and ambitions - probably achieving more than you would have done if you hadn't have known?

Unless you are actually in this position, it's doubtful whether you can answer these questions with certainty. But for hundreds of thousands of people, these decisions become unavoidable as a result of unforeseen problems within their bodies - diseases not caused by bacteria, parasites, or any other obvious outside force, but by the malfunctioning of the very substance that creates and controls us - DNA.

Genetic diseases can either be passed on from one generation to the next or they can arise spontaneously, at any time, in any one of us. The subject is now so huge and there are so many diseases that can be attributed to, or at least linked in some way with our genes, that it would take an entire book to give an overall picture. The bewildering amount of information available, most of it in a technical jargon incomprehensible to most folks, may be behind the lack of public knowledge and even interest. But if we take the time to understand some of the important issues involved, we will be in a better position to influence the decisions that affect us all, now and in the future, rather than leaving it all to the politicians and scientists.

The Story so Far

In the previous few chapters we have marvelled at how far scientists have progressed with their new toy - the ultimate force of life itself, DNA. Not content with merely understanding how they work, man has actually started to manipulate genes themselves. He has begun to 'play God' by swapping genes between animals and plants, and even between man and other animals. We seem to be so advanced with this new science of *genetic engineering* that we must surely be well on the way to solving the various puzzles of genetic disease?

With many disorders this is true - breakthroughs have been made in some of the more common inherited illnesses - but man is a long way off being in complete control of his DNA. Scientists are still far from knowing what all the genes on our chromosomes do. And there are many diseases that are due to a genetic malfunctioning, the nature of which we do not understand. In truth, we are only beginning to understand the problems and to touch on possible cures.

For most, or indeed all of our lives, we are completely oblivious to the non-stop activities of our cells - let alone the genes that control each of them. This is because the vast majority of genes usually work perfectly well, successfully co-ordinating the cell's protein production, waste disposal, energy production and even death[1]. In a perfect body, all of the cells co-operate with each other - there can be no feuding families or neighbourly disputes in the world of the human body. The end result of this harmonious communication and co-operation is a magnificent fully functional human being.

But what happens when one or more genes go wrong? This is when we may start to sit up and take notice. A problem gene is usually an ordinary gene that has changed in some way so that it doesn't give out the correct orders. If correct orders are not given, the entire cell may be affected. And if just one cell, or type of cell, starts to go awry, there can be a very significant effect on the body. Even a tiny change in the DNA of a gene may produce catastrophic results in the individual.

It is important to distinguish between inherited and non-inherited genetic diseases. An *inherited* disease is caused by a faulty gene that has been passed down from one or both parents. The children will grow up with the fault in every single cell of their bodies and may subsequently pass it on to their children. This is a major cause of heartache for many prospective parents. If you know that you or your partner is carrying a malfunctioning gene, should you test your unborn baby to see whether he or she is carrying it? If the gene *is* detected and it means that the baby is at risk of developing a distressing condition later in life, what would you do?

Part of the unpredictable nature of some of these diseases is that when you inherit a defective gene it does not necessarily mean you will develop the disease and, in many cases, the extent of the disease can vary from one person to another. For some diseases, other genes may counterbalance the fault. For others, environmental factors may have an influence.

---

[1] By a rather efficient process known as apoptosis, a cell can commit suicide neatly and without any mess. When its time has come, it breaks itself down and all the components are re-cycled and re-used by surrounding cells.

Attack of the Genetically Engineered Tomatoes

JIMMY WAS A PROBLEM GENE, A GENE WITHOUT A CAUSE

Inherited diseases affect a relatively small proportion of the population, but all of us can acquire *non-inherited* genetic disorders. Small faults crop up in our genes on a daily basis, in each and every one of us - however healthy or fit we think we are. An apple a day or a twice-weekly session at the gym cannot stave off what are commonly (and perhaps vaguely alarmingly) known as gene mutations. Any normal gene in any cell of the body may spontaneously change, or 'mutate', and start to produce effects that harm us.

Another time for potential problems is whenever your body creates a new cell. Every time a new cell is made, the entire set of chromosomes in the old cell has to be copied, ready for insertion into the new cell. With so many hundreds of thousands of genes, it is inevitable that mistakes will occur in the copying process. New cells are being produced in our bodies all the time to replace the old ones, and mistakes are being made in the copying process all the time. With each copy, more and more faults can occur in the genes. As we get older, these mistakes accumulate and the risk of problems increases.

Spontaneously mutating DNA and the accumulation of gene

defects certainly looks worrying. Who would have thought that errors are constantly occurring in this substance that our whole lives are based on, and as many as 10,000 errors every day? But fortunately for us, DNA is a marvellous molecule. It hasn't survived millions of years and created thousands of different life forms without developing some means of self-preservation. In fact, there is a miraculous system of 'proof reading' which looks for errors in newly created chromosomes, and there are special genes whose job it is to repair faults that crop up. Unfortunately, these repair genes themselves can also go wrong, or they may not be able to cope with the amount of DNA damage. And then we start to get problems. This is typically the case with cancer. A fault in just one gene can lead to uncontrollable cell growth, forming a mass of out-of-control tissue, commonly known as a tumour.

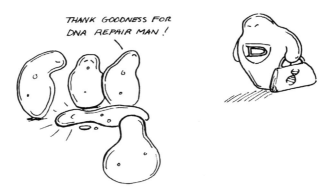

As well as naturally occurring mutations in our bodies, there are substances in the environment which are known to increase the rate of gene mutations. The most common mutagen of this century must surely be tobacco. Despite the explicit warnings on cigarette packets, there are still hundreds of thousands of people on the register of lung cancer deaths - the most obviously attributable cancer. It could be argued that some deaths from cancer can be *blamed* on something. Although they are not directly infectious agents, cigarettes, radiation and sun burn, for example, can be just as deadly as cholera or malaria. But some deaths from cancer have no apparent trigger. And it seems to be more difficult to cope with the death of a loved one when the blame cannot be attributed to anything - a case of breast cancer in a middle-aged woman who had never smoked or touched alcohol, had always lived carefully and never indulged in any of life's vices, for instance.

Breast cancer is, in fact, an example of one of the areas of genetic disease which is only understood up to a point. In fact, there are at least two faulty breast cancer genes that can be inherited, but there are also unspecified events that can occur during a lifetime - flicking a switch that causes a cancerous malfunction.

The study of human genetic disease is not a straightforward science. But as we discover more about human genes, so we can understand what happens when they go wrong. In an ideal future, greater understanding of these diseases would enable 'gene therapy'. This would be the ultimate treatment - fixing the root cause of the problem by actually manipulating the genes to cure the patient. In the meantime, we can all try and be aware of genetic problems. As a step towards an increased general knowledge, an information kiosk called the Gene Shop opened at the beginning of 1997, in Manchester Airport. This unique facility is aimed at providing user friendly education on genetic diseases and issues, to anyone who passes by!

The rest of this chapter discusses two examples of inherited genetic disorders in detail, reflecting the limited insight I have gained from knowing friends and their families who have suffered as a result of a few microscopic changes in their DNA. The following chapter casts an eye over the broad spectrum of disease known as cancer, a disease of many varieties which may arise spontaneously or may be inherited.

Cystic Fibrosis

For most of us our 30th birthday is a milestone in our lives. It marks the end of our carefree twenties, and we may revel in our new maturity or feel slightly depressed at the passing of our youth. For those among us who suffer from cystic fibrosis, the 30th birthday holds a completely different significance. It is a year of age which the statistics say will be celebrated rarely and, for those who do survive it, death is already overdue.

For the parents of children with cystic fibrosis there is a near certainty that their offspring will die before they do. For the children themselves there is a lifelong struggle for a normal life. It is a battle against not only the pain and endless infections, but also the statistics.

Cystic fibrosis is the most commonly inherited fatal disease in Caucasians, affecting around one in 2500 children. Roughly speaking, one child is born with cystic fibrosis every day, and there is no cure. It is usually diagnosed in babies or children of up to a few years old who may be having problems with recurrent lung infections or who fail to gain weight properly. Problems with the lungs are a typical feature of the condition throughout the life. Due to a fault in the cells that line them, thick sticky mucus builds up in the small passages and air spaces of the lungs, making breathing more difficult and encouraging viral and bacterial infections. Another problem occurs in the cells of the pancreas. When these go wrong, the channels in the pancreas get blocked with mucus. These channels are vital for carrying enzymes to the intestines for the digestion of food. In the early part of this century, when the disease was only just being recognised, children with cystic fibrosis used to die from malnutrition because they couldn't absorb the nutrients in their food properly. As the channels get blocked up with enzymes, the pancreas becomes inflamed and this leads to the formation of cysts and fibrosis (thickening of tissue), hence the name of the condition.

To a certain extent physiotherapy can help to clear the chest, and regular antibiotics can help stop chest infections. Special diets and enzyme supplements may also help to combat the deficiencies of the pancreas. Ultimately, however, the body cannot cope and eventually the build-up of mucus and repeated infections defeat it.

It is both intriguing and horrifying to think that such considerable problems, and eventually death, are caused by a tiny fault in just one gene - out of the hundreds of thousands of genes in a cell. It was as recently as 1989 that researchers in the USA discovered the precise location of what is known as the 'cystic fibrosis gene'. In healthy people, the gene in question usually makes a large protein, consisting of approximately 1480 amino acids (the building blocks of proteins). This protein acts as a channel for salt and water in the membranes of cells lining the lungs, pancreas and skin for example, and the protein is therefore known as the cystic fibrosis transmembrane conductance regulator, or CFTR for short. When the CFTR gene is defective, it does not make the CFTR protein properly. Hence the transport of salt in and out of the cell is disrupted, leading to the build-up of mucus in the lungs and pancreas. In addition to lung infections, parents may notice that a baby with cystic fibrosis has very salty skin.

In most of the people who suffer from cystic fibrosis, a very small section of the gene which makes CFTR is missing. In fact, just 3 letters of the gene's code, which consists of hundreds of letters, are missing.

These 3 letters would usually make the amino acid, phenylalanine. When this *one* amino acid is not made, the rest of the entire CFTR protein is disrupted and cannot function properly.

The most horrible aspect of this disorder is the unexpectedness of it. For two parents who give birth to a healthy-looking son or daughter there can be no happier moment. The baby may have no obvious physical deformities or handicaps, and it is only when cystic fibrosis is detected months or years later that the blow comes. To make matters worse, both parents must live with the knowledge that they have both passed on a faulty copy of the gene to their child. If only one parent passes on a malfunctioning CFTR gene, the child will be fine because the healthy copy from the other parent counteracts the ill effects. Each person has 2 copies of every gene in their own body and passes on one copy to the child. In order for cystic fibrosis to manifest itself, a baby must inherit two faulty cystic fibrosis genes, one from each parent. If a

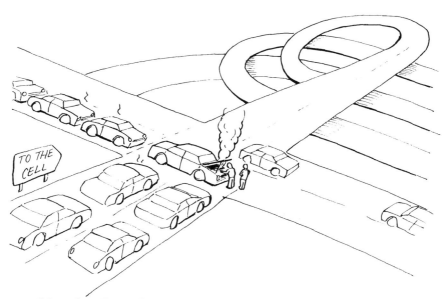

JUST ONE FAULTY GENE CAN CAUSE A BREAKDOWN IN THE CELL'S ACTIVITY

person inherits one healthy gene (from one parent) and one faulty gene (from the other parent) he or she will never develop any symptoms of cystic fibrosis. That is why most of the *carriers* of a faulty gene are unaware that they do have it.

In fact, approximately 1 in 25 Caucasians carry a faulty copy of the gene and run the risk of passing it on to their children. This is equivalent to an estimated 2 million people in Britain and 8 million in the USA. For every couple, there is around a 1 in 600 chance that they will both be carriers of the cystic fibrosis gene. If both parents carry a copy of the faulty gene, the couple is termed 'high-risk'. If they decide to have children, then the laws of probability dictate that out of every 4 children they have, one will suffer from cystic fibrosis and 3 will be fine, although 2 of these will carry the faulty gene and may pass it on to their children.

Approximately 2 years after the CFTR gene was identified, the first genetic test was brought out. Now there are several on the market which can detect whether you carry a faulty cystic fibrosis gene, and only a simple mouthwash sample is required. These tests, however, can only detect around 80 to 90% of carriers - over 200 mutations of the gene have been detected and no test can find all of these. The most common variation in the gene, however, accounts for up to 80% of cases. Because the test cannot detect all the different types of mutation, it means that even if you get a negative result, you cannot be absolutely sure that you are not a carrier of a cystic fibrosis gene.

Anybody in the UK can undergo testing or 'screening' to see whether they are carriers of the faulty gene. If you are found to be a carrier it has absolutely no bearing on your own health. You will always be healthy because you will always have a healthy copy of the gene. It does have implications though if you want to have children, in which case it may also be advisable for your partner to be tested. If he or she is also found to be positive you are a high-risk couple and will obtain counselling about the risks for your children.

Most of the babies that are born with cystic fibrosis are born to unsuspecting couples - people who never even gave the condition a thought when they were expecting a child. The fact that many of the population are carriers of a faulty gene has prompted calls for routine screening to be offered to all couples before pregnancy. This would at least be able to prepare would-be parents for any difficulties that may lie ahead. In 2 antenatal clinics in Edinburgh, screening has been available since 1992 and has proved to be very popular. Around 70% of those offered took up the offer of screening to see whether they carried the cystic fibrosis gene. Out of 22 couples identified as being high-risk, 20 opted for a prenatal diagnosis and 8 pregnancies were found to involve a child with cystic fibrosis.

Prenatal diagnosis is the testing of the unborn fetus, and can detect whether the child has 2 faulty copies of the gene and will therefore develop the condition. This detection, however, brings with it new dilemmas. The knowledge that genetic screening provides gives couples a new, heartbreaking choice. Suddenly there is a need for a life and death decision.

If you knew the destiny that faced the child you carried, would you prefer to spare it certain suffering, or would you want it to at least have a chance of life?

Out of the 8 pregnancies in the Edinburgh study where the fetuses were identified as having cystic fibrosis, *all* were terminated. It is for no one to judge in this instance what is right and wrong. What is most important is for people to fully understand what the 'genetics' mean in real terms. Unfortunately, the evidence points to a rather low long term understanding of test results. In a follow-up of 746 patients from 6 UK centres, patients who had received an explanation of their test results approximately 3 years previously were contacted and re-questioned. Of those who had been found to be a carrier of one copy of the faulty gene, 16% thought they were only *likely* to be a carrier and that it was not actually proven. Conversely, 50% of those who tested negative wrongly believed that the result meant they were *definitely* not a carrier.

Since the gene was discovered, the quest for gene therapy has been pursued fervently. The prospect of significant financial rewards for the team which is first past the post has spurred on a number of major players in the pharmaceutical industry. The race is on to be able to deliver healthy CFTR genes into the lung cells, enabling the healthy CFTR protein to be produced. The transfer of the healthy gene is via a 'vector', and experiments have already taken place with a number of different types of vector. The adenovirus, for example, is usually responsible for the common cold, but it can be modified so that it doesn't cause infection, while at the same time carrying healthy CFTR genes. The virus is introduced into the lungs and easily infects the cells. When the virus infects a cell, the genes it contains are taken up into the rest of the cell's genes. The problem with the virus is that it causes irritation and inflammation.

An alternative is to use microscopic fat globules called liposomes. These can easily fuse with the fatty wall of cells, and can release the genes they are carrying inside the cell. Liposomes have been used in human trials in the UK which experimented with cells lining the nose. The nasal lining is very similar to the lining of the lungs and is easier to reach for research purposes. The problem with liposomes is that the delivery of DNA is not very targeted. The contents are merely released into the cells, but are not directly taken up into the cell's own DNA. Researchers are now trying to develop a vector which combines the best of both technologies.

Although gene therapy does offer hope it cannot be considered a cure at the moment. There are still many basic problems. Because all of our cells are continually being replaced, a delivery of healthy genes to replace dud ones would be required every few months. There is also the problem of correcting the pancreas. This is difficult to reach and it is thought unlikely that gene therapy will ever be an option for this region, and enzyme supplements will probably always be necessary.

The ultimate solution for cystic fibrosis, and indeed any genetic disorder, would be to completely repair the faulty genes in just one step, and permanently. To be able to fix the DNA of all cells, this correction would have to take place at a very early stage in the development of the fetus. If at the single cell stage or very close, we could correct the CFTR gene, then the baby would grow with all cells containing the corrected gene. In effect, the disease would not appear in the individual and would never again be passed on. These kinds of experiments have already been tried in animals - and with some success. Using an adenovirus to carry the corrected gene, a group of mice were successfully cured of cystic fibrosis. The adenovirus was injected into the fluid surrounding the fetuses - which then absorbed them and the genes into their developing cells. The new-born animals were subsequently found to have developed into healthy creatures, and did not develop the disease.

Although these sorts of experiments have been performed in animals, there are no plans to attempt them in humans. The techniques are not nearly advanced enough and the risks are too great - not to mention the ethical dilemmas of fiddling about with the genes of a barely conceived human being. At the moment we have to accept that this most ambitious of genetic solutions is not yet within reach.

Muscular Dystrophy

Is it easier to live life not knowing when, where or how death is finally going to visit us? Or would it be better to know all of the precise details - the location, date and even time? *Knowing* would certainly help you to focus on all those things you knew you wanted out of life, and the happiness.

But suppose that you knew what your fate was because you had seen it, but you weren't quite sure of the exact details.

It might be worse than you had seen. Or not quite as bad.

It might occur later in life that you had seen. Or earlier.

Or you may cling to the hope that perhaps you can thwart this fate, that it won't happen to you after all.

Myotonic muscular dystrophy is a muscle wasting disease. It is one of a number of types of muscular dystrophy, but is possibly the most frequently inherited muscular disorder in adults. Those affected may not be aware that they have it until quite late in life. Their parents may never have experienced any obvious symptoms and may never have had reason to suspect anything out of the ordinary, certainly not a genetic disease. It may even take several years before it is finally diagnosed for certain with a genetic test, but even then the future progression of the disorder cannot be predicted with any certainty.

This disease does, however, have a profound effect - not only on the person who finds his or her life turned upside down, but also on all the relatives. By the time myotonic dystrophy has been diagnosed, the individual may have had a family and unknowingly passed on the faulty gene. Over a number of years, the sons and daughters can be witness to a gradual change in their parent. There is the slow onset of disability and ultimately death - a fate that they too may one day suffer.

Many types of muscles are affected including the legs, hands, face, neck, jaw and even intestines. The age of onset varies from person to person, as does the severity of the disease, and it is impossible to predict how badly the disorder will progress and how fast. It may first become noticeable as a small momentary stiffening of the tongue - an embarrassment at dinner - or an inability to relax your grip immediately - a painful few extra seconds after a handshake at a job interview. In time it may gradually develop into a far more significant problem -

a weakness in your arms rendering you unable to do your job, or a frustrating inability to use your legs, leaving you restricted to a wheelchair.

As with cystic fibrosis, the gene in question has been successfully identified. It is known as the myotonic dystrophy protein kinase (DMPK) gene and resides on Chromosome 19. But there is a special twist to this genetic disorder. If you inherit just one faulty copy of the gene, you are almost certain to develop the disease. This is because the DMPK gene is what is known as a dominant gene. Even if you inherit a healthy copy from one parent, the faulty copy inherited from the other parent will always override it.

The error that occurs in the myotonic dystrophy gene is also somewhat different to the relatively simple fault in the cystic fibrosis gene. Rather than undergoing a single mutation, the gene in question multiplies itself many times. This means that where there should be just a single copy of the gene in the chromosome, there arise many copies. These sections are termed 'unstable' regions of DNA; they make it much more difficult for the natural proof reading mechanism in our genes to do its job accurately, and they also have a tendency just to get bigger and bigger, consuming a larger part of the chromosome.

As the faulty genetic segment is passed on from one generation to the next, it increases in size and disrupts the normal functioning of the other genes even more. Once there are 50 or more copies of the repeated gene in the chromosome, a threshold is crossed and suddenly the harmful, muscle wasting effects of the disorder become apparent. When the segment gets larger, as it is passed on to a son or daughter, the symptoms become more severe. Respiratory problems may develop and there may even be mental retardation.

There is a genetic test available for this condition but there is a question mark over how it should be used. Certainly, for individuals who develop muscle weakness symptoms, the test can be used to help diagnose the condition for certain. But prenatal diagnosis in the unborn is a much more sensitive issue. Can the choice of life and death ever be ethically made for this condition? On the one hand, there is no cure and little symptomatic relief for myotonic dystrophy. But on the other, it is impossible to predict how severely your child will be affected, and how many years of life it will enjoy before muscle problems will begin. Would screening merely place feelings of helplessness on parents, causing unnecessary suffering? Or could it be beneficial? It is possible that many couples would rather not know -

children who have myotonic dystrophy, or any other genetic disease, are as loved and cherished as any other. It's just not an issue. And for many people, abortion would simply not be an option anyway.

But there may still be benefits to screening. It could be argued that if the faulty gene was detected it could help prepare couples for problems which lay ahead, and appropriate counselling could be given to affected children. It could also increase awareness in the whole family, and may be important for other relatives who want to have children. There is no easy solution to these problems, but increasing our awareness and understanding can only help.

Summary

Technology has progressed in leaps and bounds to enable us to reach inside a cell, to probe into its nucleus and analyse deficiencies in the DNA which lies therein. We can actually pinpoint problems down to individual genes that lie in the huge cosmos of our chromosomes, and we can perform genetic tests to identify who carries a faulty gene. We've briefly looked at the possible advantages and disadvantages for the families who are immediately affected. But what does this new-found technology mean for society at large? The potential now exists for the information gained from screening to be used against individuals. It is essential that this does not happen. Employers must not be allowed to discriminate against people who carry faulty genes. Life and health insurance companies must not have access to the results of genetic tests. And we should never be instructed to have such tests for the purposes of insurance.

This unfortunately does not seem to be the way that things are progressing, however. In the UK, if we do have a genetic test voluntarily, we are not permitted to keep the results to ourselves when it comes to insurance. In February 1997, the Association of British Insurers decided that anyone applying for life insurance must reveal the results of any genetic test that he or she had taken. Many insurance companies proposed to ignore the results unless the application was for insurance of over £100,000, linked to a new mortgage for example. But this still potentially affects hundreds of thousands of people, and it may put people off having a genetic test which would benefit them. In addition, the legality of holding results that are not to be used was questioned recently; it may breach the Data Protection Act, 1984, and the Association has been made to reconsider its policy. But it is likely that insurance companies will still use the results for applications concerning large sums.

A poll was commissioned by a group called the Genetics Forum in 1997 to examine the public's attitude to gene testing. Out of 1,000 people questioned, more than 25% said that they would not take a genetic test if, as a consequence, they would have to disclose the results to an insurance company. But apart from the fact that many people would be put off genetic testing if it meant that insurance premiums would be substantially increased, or if they experienced greater difficulty in finding a job, this kind of discrimination is fundamentally unethical. It is as bad as discrimination against people based on their race, sex or hair colour, because it looks at something that we are born with. Our genetic material cannot be changed - we have to live with it, and should not have to pay any kind of penalty for what it contains. Once society has realised this crucial point, then it can begin to address the practical issues and costs involved in using screening to our advantage.

If screening were to become a matter of routine, as commonplace as measles or polio vaccination, who should be responsible for its co-ordination? Who would organise public awareness and education programmes about genetic disorders and their treatments? There should be proper supportive measures in place to help and advise people once they have been screened. People who were not expecting to find they were carrying a problem gene would be plunged into worry following a positive result, and patients who discover genetic problems will be placed in a difficult position - should they have to tell their partners and relatives who also might be carrying the gene?

This chapter has also touched on the possibilities of gene therapy. This is the ultimate solution to all genetic disorders. To insert a healthy copy of a defective gene into a patient's body is fighting the disease at the most fundamental level. Rather than treating the symptoms, the root cause of the problem is dealt with. In the UK, the Human Genetics Commission was set up to review the social, economic and ethical implications of screening and gene therapy trials, and many experiments in gene therapy have already begun. There are also various safeguards in place to ensure that all work is of a satisfactory standard and justifiable; protocols for gene therapy trials in humans have to be approved by the Gene Therapy Advisory Committee as well as the Medicines Control Agency. For genetic screening, the government-appointed body, the Advisory Committee on Genetic Testing, proposed towards the end of 1997 that companies planning to launch commercial genetic tests must also submit their tests for approval. In addition, they also proposed a code of practice for companies, and recommendations that the tests on sale directly to the public should be limited to disorders like cystic fibrosis, rather than more complex ones like Alzheimer's disease.

There is, of course, still a long way to go. Many genetic diseases are still much of a mystery and, even if we could find answers to all the genetic problems we have now, new mutations in the future will quickly alter genes and create yet more riddles. The next chapter looks at one of the biggest killers - cancer. This is again a type of disease which has genetic problems at its heart, but problems which are extremely varied and complex.

6

KILLER IN US ALL

'At West London yesterday Emily King, 39, a widow, of Aintree-street, Fulham, was charged with attempting to murder her son Herbert, aged seven years, by cutting his throat with a razor, and with attempting to commit suicide by cutting her own throat.

It was alleged that the prisoner was driven to desperation by poverty and hunger. The police surgeon stated that the woman had a wound in the throat 2in. long and the boy had a wound five or six inches in length. The wound in each case extended to the bone.'
The Times, Thursday, September 22, 1910.

Nowadays such a shocking tale would surely make the headlines just as easily as it did over 80 years ago. This is the tragic story of a mother who felt compelled to take the life of her young child before turning the blade on herself. Although it was a deliberate attempt at murder and suicide, we instinctively feel sorry for the woman. She must have been driven to this insane act by an unbearable hardship. Society had let her down and this was her solution.

How many of us would have the nerve to murder another person, let alone our own child? And who among us could imagine raising a knife to their own throats before cutting - knowing the intense and possibly lingering pain that would have to be suffered before death? Perhaps it is this image of an acute agony in a bloody mess that is so repelling.

Maybe it is the slow, chronic nature of cancer - an invisible death within the body - that makes it more acceptable. For it seems that despite the warnings that cry out from every cigarette packet and advertisement, hundreds of thousands of people persist in harming themselves and their children through smoking every day. And it is in the 'lower' classes where smoking is more prevalent and the deaths from cancers of the lungs and stomach, for example, are the highest.

By putting a cigarette to our lips are we not just as guilty of suicide as we would be by raising a knife to our throats? By smoking in the presence of others, including our children, are we not just as guilty of murder? Cancer takes the lives of around 165,000 individuals in the UK and half a million in the US every year. It is estimated that around 1 in 3 of us will probably develop some form of cancer in our lifetime. Although any part of the body can turn cancerous, the most commonly affected areas are the lungs, breast, skin, prostate and colon. Deaths from lung cancer are the most common and as many as 90% of cases are said to be directly related to smoking. And then there are all the deaths from cases of mouth, throat, stomach, pancreas, oesophagus, kidney and bladder cancer, among others, which are caused by smoking. Perhaps not surprisingly, most of the deaths from cancer are thought to be preventable.

The death rate in middle aged men is three times as high in smokers of cigarettes as it is in non-smokers. In an in-depth study of life expectancy in 7,000 British men, it was found that approximately 42% of lifelong smokers would live to see 73 years of age, compared with nearly 80% of lifelong *non*-smokers. Tobacco giants have traditionally fought the scientific evidence and denied that their products are addictive, but the evidence and arguments against them are ever increasing. More individuals are pressing for damages and even entire states within the USA are taking up lawsuits in an effort to recover money spent on treating people with smoking related illnesses. In 1996 a significant milestone was passed when a 66-year-old man who had smoked Lucky Strike cigarettes for 44 years and had developed lung cancer was awarded $750,000 (£500,000) in a Florida court. And even more recently, following a flurry of lawsuits, tobacco companies have started to co-operate with individual states, and to admit some kind of responsibility. In the US at least, tobacco companies have negotiated a deal in which they gain protection from further legal action, in return for a fine of nearly $370 billion (equivalent to £230 billion) and various other concessions such as admitting that cigarettes are addictive and cause cancer. Some people feel that this is not nearly enough; the fine is to be paid over a number of years, and could easily be found by increasing the price of cigarettes slightly. And some argue that only a complete ban on smoking will be satisfactory - surely they argue, tobacco companies are just as bad as drug pushers, maybe plying an even deadlier trade?

In 1997, the British Medical Journal published an open letter from J. Asvall, Regional Director of the World Health Organisation, to the heads of the 51 European member states. In this he states that there are 1.2 million deaths every year from tobacco related diseases, and that there will be 2 million per annum by the year 2020. This will be equivalent to a fifth of all deaths. And he urges all governments to look closely at what they can do to turn the tide.

It is hard to believe that smoking has not yet been banned from bars and restaurants when it is a well known fact that non-smokers who are subjected to cigarette smoke are also at risk. Waking up in the morning with clothes and hair that wreak of a repugnant odour is bad enough, without also having to entertain the thought that you may have seriously damaged your health. It is estimated that over 50,000 deaths in the US every year are associated with passive smoking, and non-smokers who live with smokers may have a 30% increased risk of death from heart disease or heart attack. There are thousands of chemicals in cigarette smoke and at least 40 cancer-causing agents, or carcinogens. Researchers in the USA performed experiments in 10 male non-smokers to see whether a known carcinogen in the smoke was absorbed by passive smoking. The men were exposed to conditions which mimicked a heavily smoke-polluted bar for an hour and a half on two separate occasions. The men's urine was collected for 24 hours before and after their exposure and tested for a substance called 'NNK' (4-(methylnitrosamino)-1-(3-pyridyl)-1-butanone). NNK is known to be highly potent in causing lung cancers in rodents, in particular, a type of cancer more commonly found in passive smokers. After collecting the men's urine and analysing it, the researchers found that all of the men had absorbed this carcinogen.

One of the most frightening forms of passive smoking is the transfer from a pregnant mother into her unborn baby. The umbilical cord which links the mother and fetus carries blood, which passes on food and nutrients to the growing child. Because this blood supply is inextricably linked with the mother's own blood network, all the substances that she absorbs are fed directly to the child, and that includes the chemicals and carcinogens of inhaled cigarette smoke. One of the ways that the poisonous chemicals in cigarette smoke can damage the fetus is by reducing the circulating oxygen in the blood; if insufficient oxygen reaches the brain of a developing baby, crucial nerve pathways may be affected, as well as other vital tissue developments. In a US study in 177 boys aged 7 to 12 years old, it was found that conduct disorders were significantly more likely in boys born to mothers who smoked more than 10 cigarettes per day during pregnancy, suggesting that brain development was indeed affected.

By developing a technique to dye the DNA in cells, researchers in California were able to go one step further, by studying a certain type of mutation that occurs in chromosomes. New-born babies that were born to smoking mothers had 50% more mutations in their cells compared with babies born to non-smokers. No one knows, however, what exactly the cancer risks are for these babies.

Studies in men have also shown that smokers can pass on genetic mutations in their sperm. Researchers at the University of North Carolina studied 15,000 children and found that children whose fathers smoked more than 20 cigarettes per day were twice as likely to suffer from birth defects such as harelip. In a second study, the histories of 220 young children who were suffering from cancer were analysed. Leukaemia and cancer of the lymph nodes were found to be twice as common in children whose fathers had smoked in the year prior to the child being born.

The damage that is passed on in a father's sperm is not only to his own son or daughter, but also to all of *their* children. Because the damage is present in the fundamental genetic material, it then becomes part of the genes that are passed on throughout every subsequent generation. Fathers, and indeed mothers, who smoke are therefore not only responsible for damaging their own genes, but also the genes of each of their children and their children's children, *ad infinitum*.

It is never too late to stop smoking. The damage to your life expectancy can be reduced even if you are in your 50s. In 1996, in an effort to encourage its residents to quit the habit, Guernsey became the first place in the UK to approve a complete ban on tobacco advertising. Local newspapers, hoardings and sports events were all instructed to be free of advertising and the minimum age for buying tobacco increased from 16 to 18. Central government also seems to be waking up to the need to tackle the costly issue of cancer, including smoking. In the UK, the former Conservative government proposed a *Health of the Nation* strategy. This included the aims to reduce ill heath and death caused by breast and cervical cancer, to reduce ill health and death caused by skin cancer by increasing awareness of the dangers of ultraviolet light, and to reduce ill health and death caused by lung cancer and other conditions associated with tobacco use, by reducing smoking and tobacco consumption. In order to meet the third objective, the strategy specified that overall cigarette consumption should fall by at least 40% by the year 2000, and at least a third of women smokers should stop smoking at the start of their pregnancy. It is this focus on prevention rather than treatment which may in the long term have the greatest impact on reducing the number of deaths. It needs every individual to take a careful look at his or her lifestyle and make a conscious decision about their own health.

There have been successes in curing some forms of cancer, such as acute lymphocytic leukaemia in children and testicular cancer, by means of drugs and radiotherapy. And a number of techniques have also been developed to detect cancers earlier, notably breast, colon and cervical cancer. If cancer is caught earlier, there is more chance of treating it successfully. The problem with finding a definitive cure for cancer is that the disease is so complex and there is no single, precise fault that can be easily fixed. It begins as a malfunctioning in one or a few healthy cells, and gradually there is a breakdown in communications with surrounding cells. The normal activity that is essential for a well organised, working relationship is lost, and the rogue cells assume megalomaniacal personas, beginning to divide and multiply out of control. The resulting mass of unwanted tissue physically disrupts the other, normal cells which are trying to get on with their jobs, and whatever part of the body is affected ceases to function properly. Mutant cells may break off from the original group and travel to other parts of the body, forming new settlements and centres for malignant growth, called metastases. The whole process can take many years, but may only become noticeable in the later stages, when advanced tumours have already taken hold and are beyond effective treatment.

Huge advances in our knowledge about the intricacies of cells have helped scientists to learn why a normal cell may turn into a cancerous cell. It seems that one or more genes in the cell undergoes a change, or *mutation*. This brings about a complex series of interactions and perhaps more mutations, which ultimately incite the cell to divide and proliferate out of control. Oncogenes are essential types of genes which are needed to regulate the lifespan of cells, including their timely division. If they go wrong, the cell's ability to divide sensibly and multiply at the correct time becomes seriously compromised. Nearly all cells contain oncogenes which means that *every* part of the body has the potential to turn cancerous if they go awry.

Another type of gene that has been linked to many forms of cancer is commonly known as *p53*. This is what is known as a tumour suppressor gene, and approximately 60% of all cases of lung cancer are thought to be caused by a spontaneous change in it. Recent studies have even shown that a particular compound in tobacco smoke physically binds to *p53* in cells and damages it at a number of places along its length. When it is working normally, *p53* acts as a kind of braking system for cells. It provides a natural anti-cancer mechanism which ensures the cell does not divide and proliferate when it should not. This means that when *p53* develops a fault, the cell has no mechanism for controlling its own division. Certain places along the *p53* gene appear to be more susceptible to damage than others, particularly from environmental factors such as ultraviolet radiation. In addition, the usual apparatus that the cell employs to repair any DNA damage does not appear to work very well in these areas. There now appears to be a direct link between a carcinogen in cigarette smoke, called benzo[a]pyrene, and mutations in the *p53* gene which are associated with lung cancer.

The rest of this chapter looks at two very different types of cancer in more detail, breast and skin cancer.

Breast Cancer

The majority of cancers are a result of an unpredictable change which can arise in a gene at any time during a person's lifetime. These are sometimes known as sporadic cases, and certainly the majority of lung cancers are of this type. Breast cancer is slightly different however. Although most cases do occur spontaneously in an unsuspecting individual, a significant proportion can be traced to a faulty gene that has been inherited. These people may know of others in their families who have experienced the same type of cancer, and for other close relatives there is a frightening prospect that they may also one day develop the cancer.

The shocking statistics announce that over 30,000 new cases of breast cancer (including both inherited and sporadic) are diagnosed every year in the UK, and approximately 13,000 people die annually. In the USA, the annual death toll is around 50,000. As with all cancers, these appallingly large numbers cannot reflect the true agonies that the individual and their families must endure. The disease can strike in men as well as women but, for women, the disease is probably the most unpleasant and alarming type of cancer that can befall them. Not only do they have to cope with the frightening label of 'cancer', but these women are submitted to the ravages of disease in one of the most sensitive parts of their bodies, central to their femininity. To make matters worse, breast cancer is sometimes accompanied by the development of ovarian cancer. Although this form is less common, it is more virulent and causes around 4,000 deaths every year in the UK.

Around 5 to 10% of cases of breast cancer are thought to be caused by a damaged gene that has been inherited. Most of the research into breast cancer has focused on trying to find the gene or genes responsible for the inherited form as it was hoped that these defective genes would also provide the clues to the remaining 90% of cases which arise spontaneously.

By studying families in which a number of individuals developed the disease, researchers were able to identify the first gene known to play an important role in around half of all inherited breast cancers.

The gene was first described in 1994 and is termed *BRCA1*. In its normal state this gene helps to suppress tumour growth, and in this respect it is similar to *p53* which commonly malfunctions in lung tumours. It is estimated that 40 to 85% of women who inherit a defective copy of the *BRCA1* gene will develop breast cancer at some time in their lives, and half of them will do so before the age of 50. If you have this gene, you also have the trauma of knowing that you run an increased risk of developing ovarian cancer as well.

A second breast cancer gene was identified the following year, and termed *BRCA2*. This is located on a completely different chromosome and is thought to account for 70% of inherited cases that are not due to *BRCA1*. It is also associated with an increased risk of the disease in men.

The initial excitement that greeted both discoveries has to some extent now passed. Neither of the two genes seems to play an important role in non-inherited cases of breast cancer and they have done little to elucidate the mystery of cancers which arise spontaneously in women for whom there is no family history of the disease. Another disappointment has been the finding that the *BRCA1* and *BRCA2* genes can have many different types of faults, or mutations. In contrast with cystic fibrosis for example, for which only three different mutations of the gene account for nearly 90% of people with the disease, there are over 100 different mutations of the *BRCA1* gene and any one accounts for only a small percentage of cases. This has important implications for genetic screening. Any test that is performed will never be able to detect every conceivable variation, but ideally it should be able to pick up as many forms of the faulty genes as possible. Any woman who is given a test to detect whether she carries a breast cancer gene must be made aware that only a *positive* test result is relatively informative. A negative test does not guarantee that she is free from a faulty gene. There is also confusion about what a positive result means in a woman who has no relatives with breast cancer. Because the defective genes have only been found to play an important part in inherited cancers it is difficult to say what effect, if any, they will have in people with no family history of breast cancer.

The decisions that genetic screening bring with them are also hard to cope with. For some women the removal of the breasts (mastectomy) and ovaries (oophorectomy) is a small price to pay to help insure against death. Such drastic surgery has previously been only the most radical resort when advanced cancer has been diagnosed, not the *preventative* action of a woman who *may* develop breast cancer in the future.

Mammograms offer a physical screening which looks for actual cancerous growth and are now essential for detecting breast cancer in the general population. Of 6,500 women over the age of 50 who were found to have breast cancer in the UK in 1994/95, approximately 4,000 were found to have cancers that were less than 15 mm or were at the pre-invasive stage. It is estimated that screening for breast cancer now saves over 1,000 lives every year. This form of screening, however, is completely different to the world of genetic screening, which looks at the *potential* for cancerous growth based on your genes. Would you consider that 'an 85%' risk for developing breast cancer was sufficient for having both breasts and ovaries removed?

The good news is that the race to develop some form of gene therapy for cancer, including breast cancers, is gaining speed. Although the developments have probably been rather over hyped by the media, there have been breakthroughs, and researchers are learning better techniques all the time. The adenovirus is a type of virus that frequently invades our bodies and gives us the common cold. This ability to infect us has been put to good use by researchers who have used the virus in gene therapy experiments, for all sorts of genetic diseases. It has been found that the injection of an adenovirus, which has been genetically engineered to carry useful genes, has a striking effect in mice. Special strains of mice were engineered to develop human breast cancer so that they could mimic the disease in humans. They were then injected with an adenovirus that had been genetically engineered to carry copies of the human interferon (IFN) consensus gene. Following the injection, all of the tumours were found to have shrunk or disappeared completely.

The direct injection of healthy copies of genes has also been found to be beneficial. Tumours in lung cancer patients who were given healthy genes all stopped growing or shrank. Even the shooting of DNA-tipped gold bullets directly into tumours has been attempted. And at the end of 1997, an Israeli company, called Medical Targeting Recognition Technologies, announced interesting results using bullets tipped with a genetically engineered hormone that had been fused with a modified toxin, usually produced by bacteria. The hormone was able to bind to the cancer cells, and the toxin killed them. Using experimental mice, the bullets were fired into the tumours and were found to be very effective.

One of the most recent, and rather puzzling discoveries, has been the 'bystander' effect. In treating a selected area of cancer cells, researchers have found that the surrounding cancerous cells also appear to be 'treated'. Some kind of message may be transferred between the cells, and this could be useful in all kinds of anti-cancer therapy. The researchers always stress, however, that there is a cautionary note for all of these encouraging finds. Yes, there are advances, but we are still a long way from a satisfactory treatment for all.

Skin Cancer

The skin is our first major line of defence against the marauding forces of the outside world, keeping out infections and protecting our delicate internal organs from every day damage. In the course of duty it suffers all kinds of injury but fortunately our bodies can usually cope, ensuring that all cuts and bruises are healed and back to normal within a few days. The skin has even evolved a mechanism for protecting us against the harsh ultraviolet rays of the sun by tanning. This lovely golden brown colour that signifies our skin is producing melanin has now become a fashion accessory and, rather ironically, a symbol of health and beauty.

The large wart-like or ulcerated growths, or the bleeding scabby moles that are skin cancer, are far from attractive and certainly not healthy - as can be testified by the 1,000 or more people in the UK who die every year from them.

Skin cancer can begin in pigmented cells such as moles (melanoma) or other skin cells (non-melanoma). The highest mortality from skin cancer can be blamed on the malignant melanoma variety, and 80% of these cases are thought to be caused by exposure to the sun. This is not to say that skin cancer does not arise in other areas. Areas of the skin which are never or very rarely exposed to the sun, like the base of the foot for example, can also give rise to cancerous growths. These may even be the most dangerous places because people least suspect that the pimples or painless lumps that appear here are of any harm. They may let them grow without concern for months or even years without realising that they are just as dangerous as a mole on the back of the hand that suddenly changes.

The precise nature of the relationship between skin cancer and ultraviolet (UV) radiation has yet to be determined although it is thought that UV-B light is probably primarily responsible. The sun's ultraviolet rays are categorised into three types depending on their wavelength. The UV-A region consists of light wavelengths from 320 to 400 nm; the UV-B region is around 290 to 320 nm; and the UV-C region is less than 290 nm. Our atmosphere and ozone are transparent to UV-A light but oxygen in the atmosphere and the ozone layer absorb UV-C so that none reaches the Earth's surface. UV-B is only partly absorbed and continuously bombards us, and with the thinning of the ozone layer, we are becoming less and less protected from all UV light. In addition, many people are unaware that some drugs make us more vulnerable to light, thus increasing our susceptibility to damage. Oral contraceptives, tricyclic antidepressants and oestrogens, for example, are all photosensitising drugs which may make us more prone to sunlight damage.

The DNA in our cells is curiously susceptible to the UV-B form of light in particular, and this is thought to be responsible for the sunburn which arises from over exposure to sunlight. But there is increasing evidence that UV-A also damages the skin. Consequently all good sunscreens now filter out both forms of light. Historically 'sun tan lotions' have been used as an aid to obtaining that perfect tan, but they are now seen more as a protective sunscreen, and have to display a Sun Protection Factor (SPF). This number indicates how much longer you can stay in the sun before you burn, so the higher the SPF, the greater the protection. The protective effects of sunscreens are, however, dependent on how sensibly they are used. If people merely use them to stay out in the sun for the longest possible time, burning at the end of the day instead of at the beginning (if they hadn't used the sunscreen), then they will have served no useful purpose.

Sunburn, often the dreadfully painful beetroot variety, is one of those things that always happens to at least somebody at the local resort on holiday. But who hasn't fallen asleep on the beach or been unaware of the ferociousness of the sun because of a cool breeze? These small incidents may cause suffering at the time but by the time you've applied the soothing 'after sun lotion' and the redness fades away, it may be too late. Individuals who suffer from melanomas are twice as likely to have experienced at least one episode of severe sunburn, and three times as likely to have had several episodes of severe sunburn, as people who do not have melanoma.

But it is not only a history of sunburn that can put you at risk. Some skin cancers, perhaps the non-melanoma varieties, may be linked with cumulative exposure rather than single episodes of severe exposure. Year on year of cautious tanning may be just as harmful as a single incident of sunburn. It has also been found that skin that is intermittently bared - on holiday once a year perhaps - is more prone to developing cancer than skin that is continuously exposed.

Recent evidence shows that tanning itself is a direct consequence of damage to DNA caused by UV light. Even exposure sufficient to turn skin slightly red or brown may be *lethal* to cells. And eventually the accumulation of genetic effects over a number of decades proves too much for the body to bear.

1. COAT WITH OIL
2. SELECT A MEDIUM/HIGH HEAT
3. TURN OCCASIONALLY UNTIL BROWN

Whatever the nature or cause of skin cancer, the most crucial factor for survival is early detection. If a tumour is left to grow, the prospects for survival diminish rapidly. For example, for tumours that are less than 0.75 mm thick, up to 99% of people can be guaranteed a 5-year survival. For individuals who have tumours that are more than 4 mm thick, less than half will live for more than 5 years. Prompt removal of the tumour at an early stage is vital. With larger tumours, more of the surrounding tissue has to be taken out to help ensure that all the cancerous cells are removed, but there is also the risk that the cancer has spread to other parts of the body and then there is no cure.

Gene therapy experiments have been attempted in this form of cancer, although it is slightly more tricky, as the skin is a huge organ which envelops the entire body. Trials in humans with malignant melanoma have attempted to treat cancerous cells with a substance called GM-CSF (granulocyte macrophage-colony stimulating factor). The treatment did appear to stabilise the disease and increase the average survival time by over a year, but only in some patients.

With all cancers, screening and gene therapy is only a small part of the battle in reducing deaths. Certainly, the knowledge that screening provides can help families with inherited cases of breast cancer for example, but the vast majority of cancers are sporadic - they arise spontaneously in individuals and, to a large extent, are preventable. By far the greatest means of reducing cancer deaths is for each of us to take responsibility for our own health, to be aware of the dangers of sun exposure or smoking, for example. There has to be a realisation that our bodies are finely tuned, living, breathing machines, which can only put up with so much.

Attack of the Genetically Engineered Tomatoes

7

THE TRANSFORMATIVE YEARS

'I'm remembering a man I saw today. He trapped me in his gaze and held me transfixed with his deep, drawling voice. He talked and I listened, but I could only take in things that didn't belong to me.
His hands.
His chest.
His mouth.
This was a handsome face, in a quirky kind of way, but there was something more. I was attracted by a whole range of indefinable looks and gestures, a slow confidence and directness in his manner, and a humour.'

We can experience such a huge range of strange and intense emotions, twist them with daydreams and memories, and live experiences in our minds to which no one else has access. And the world in our heads is without boundaries, our thoughts meander on their own volition, directing our actions and our lives. Is this uniquely human brainwork ultimately defined by our genes? And has mankind developed such profound abilities through a chance accumulation of genetic changes alone? These are surely the most interesting genetic questions of all time.

When we decide to what extent genes define us as human beings, we are effectively choosing whether to believe in science alone or a God. A number of things apparently set us aside from other animals (although we cannot be sure as we are unable to read their minds) - an imagination and the ability to daydream and fantasize, an appreciation of music and art, altruism and a moral code, a consciousness of ourselves and a variety of abstract concepts that allow us to examine the nature of life. There are many who argue that these are gifts bestowed by an unseen creator, that they are features of a 'soul' perhaps, as well as the human brain. There are others who will persuade you that these are not abstract *gifts* but are merely the electrical outputs of a finely honed brain - the result of millions of years of cumulative change, making us an extraordinary product of natural events.

The current, widely-held belief, with much evidence to back it up, is that mankind is an extremely advanced branch of an evolutionary tree that has been growing for more than 3 billion years. The main mechanism that is put forward for this evolution of all plants and animals, is based on genes. It is proposed that changing genes have created new and improved forms of flora and fauna over time, resulting in the diversity of life around today.

There is no doubt that evolution has taken place, and that genetic changes underlie it all, because genes are the ultimate blueprints for life as we know it. But, just as we have to decide whether our mental capabilities are gifts or chance complexities of nerve impulses, so we should ask ourselves, has our evolution been a result of purely random genetic changes, or has it been genetic change that was directed in some way?

The stance you decide to take will depend on how you weigh up the evidence. There is no doubt that human beings have only just appeared on the Earth, within its relatively recent history. But were they simply the next creations that were fortuitously stumbled upon in a natural course of events, or was their design penned and executed by an unimaginable force outside of the restricted confines of our universe and imaginations?

Whatever the answer, one thing is for certain. Evolution of all life on Earth has involved genetic mastering on an impressive scale. The force of nature, whatever that might be, is the most supreme genetic engineer of all time.

The Story So Far

In the midst of a particularly furious boiling temper, the type that most young adolescents seem to pass through, the 1,000-year-old Earth gave rise to microscopic cells of life. How it did this is open to speculation but, fortunately for them, these bacteria and algae actually thrived in an oxygen-deficient, rather unpleasant environment that would be unrecognisable today. As a by-product of their peculiar cellular functions, oxygen was released and began to accumulate, eventually creating a completely different atmosphere and an ozone layer on top. This was good at filtering out ultraviolet radiation from the sun, and a warm, snug planet was well on its way. Millions of years passed and a couple of bacteria decided to take the entrepreneurial step of joining together. One thing led to another and soon large numbers of cells were forming live-in communes. They were a social and co-operative lot and began to interact and work as a team. Special divisions were given different tasks to conduct by themselves, and gradually each group became more and more complex. Multicellular organisms had arrived.

Life went on much the same as usual for a while and then suddenly, *anarchy*. In a chaotic explosion of life, around 600 million years ago, huge numbers of bizarre and plainly ridiculous-looking worm and jellyfish-like creatures appeared, as if from nowhere. In the Rocky Mountains of British Colombia, Canada, near the Burgess Pass, there lies a quarry known as the Burgess Shale. The marine animals that were preserved in its mud, in exceptionally fine detail, despite their soft bodies, are evidence of a very strange world indeed. Most of the creatures, no more than a few centimetres long, bear no relation to any other alive today, and the huge number of different forms even overshadows the variety of animals that exist today. How there was such a rapid explosion of new life and why such an incredible diversity developed, we will probably never know.

Almost as suddenly as they had appeared, they vanished, with only a small number of exceptions. One of the lucky ones, an innocuous-looking, worm-like animal called *Pikaia*, possessed a distinction. It was a feature called a notochord - a stiff rod that acted as a rudimentary backbone. Could this creature be the ancestor of all vertebrates, including man? What if it had become extinct along with its fellow Burgess Shale compatriots all those millions of years ago? Nobody knows.

It was now around 450 million years ago. Time was pressing on and a more specialised marine animal was called for, something with a bit more purpose than the simple worms and jellyfish that had been taking shape. The first animals with real backbone (quite literally) were early forms of fish. These early vertebrates developed quickly and, as well as the hard backbone, they developed jaws and a more sturdy skeletal structure than their forefathers. Fish developed into two main groups - cartilaginous and bony - and members of both are still around today. Sharks and rays, for example, have skeletons made of cartilage, and bony fish - which the majority of fish species are now - have a skeleton of bone infiltrated with calcium.

Whilst all this was going on underwater, a few adventurous forms had seen the light. Early insects took to the air and plants began to appear on dry land. The greenery was still rudimentary and kept a close attachment to the water that had previously been its home. But gradually generations of newcomers realised the benefits of waterproof coverings for their leaves, thus minimising water loss, and grew roots with which to absorb water and nutrients. With roots to help anchor their structures in the soil, and woody stems to give them stability, plants such as ferns reached a new elegance, growing taller and colonising unexplored territories.

It was in the Carboniferous period, around 380 to 260 million years ago, that a select group of fishes formed a posse with the sole mission to make the move from water to land. A large number of personal sacrifices were required. Instead of gills, lungs had to be developed so that oxygen could be filtered from the air rather than from water, legs were required instead of fins, and a body structure that could support and move them around on land was necessary.

Amphibians finally completed the traumatic move from water to land, but remained from that day until this, confined to places near water for reproduction. The aquatic stage of their life cycle has not altered and - whilst the metamorphosis of egg to tadpole, and then to adult amphibian is a triumph - the next group of animals thought that they could go one step further.

I'M SORRY MUM, BUT I HAVE TO STAND ON MY OWN TWO LEGS NOW

Vegetation was now really beginning to get to grips with the opportunities of land, and spread its conifers and palm trees to as many places as would have them. Coal swamp forests and giant club mosses of up to 100 feet tall provided the early life on land with plenty of food and, with no predators around, the amphibians never had it so good. That was, until the first reptiles arrived. It must have been a huge shock. One minute amphibians had nothing to do but munch and daydream. The next, a vicious group of carnivores had appeared, quickly diversified into hundreds of different forms (including some vegetarians of course) and decided to rule the Earth for about 150 million years.

Rather than having to return to water to breed, the reptiles devised a plan to allow them to reproduce in dry, arid environments. They fertilised their eggs before they were laid, and then encased them in a hard protective, waterproof shell containing nutrients. The young could develop inside the shell and the creature that finally emerged was a miniature replica of the adult. The adults had a scaly, leathery complexion giving them the benefits of a fully weatherproof skin, and this adaptation allowed them to successfully hold down the best full-time occupation of any creature - basking in the sun for as long as possible to warm their 'cold blood'.

The reptiles were a highly successful group of animals and included those majestic monsters, the dinosaurs. From 250 to 65 million years ago, throughout the Triassic, Jurassic and Cretaceous periods of the entire Mesozoic era, these great giants enjoyed an astounding reign. They included the largest and most spectacular creatures that have ever walked the Earth, and there is evidence that they were capable of social and quite complex behaviour, even parental care.

Not content with their complete invasion of the land, they devised ways of taking to the air. Pteranodan, for example, which lived approximately 80 million years ago, had stretched skin rather than feathers on its wing-like arms, and a long beak with no teeth. Some such as ichthyosaurs, a dolphin-like creature, even went back to the sea.

And then, in another inexplicable mass extinction, about 65 million years ago, approximately two thirds of all species, including the dinosaurs, were wiped out. However destructive the human race is, we cannot compare to the forces of Mother Nature herself. It is estimated that over 95% of all species that ever existed are now extinct. Put this in context with the number of species that are around today (approximately 10 to 30 million), and you have an idea of the billions of creatures that must once have inhabited the Earth. Although separate species have died out through natural wear and tear over time - a process known as background extinction - the annihilation of great

THERE IS EVIDENCE THAT DINOSAURS WERE CAPABLE OF SOCIAL AND QUITE COMPLEX BEHAVIOUR

numbers has occurred in a number of mass extinctions throughout history. These events must have been truly catastrophic to have affected tens of thousands of species. The most likely explanation, particularly for the devastating end of the Age of the Reptiles, seems to be a global change in climate, possibly due to a collision between the Earth and a large meteor. This, it is postulated, created massive dust clouds that screened out the sun's light and heat from the Earth's surface. With such a significant drop in energy, many species found it impossible to survive.

By whatever means, this strange twist of fate set the Earth and its inhabitants on an entirely new course. Some reptiles continued to exist, and their descendants - crocodiles, snakes, lizards and turtles - still enjoy life today. Another line of reptiles also successfully evolved into birds, possibly via the Archaeopteryx, a creature with a skeleton similar to that of small dinosaurs, but with wings covered in feathers, a beak with teeth, and a long bony tail. However, the main change occurred again on land. With the old rulers deposed, it was time for a new order to assume its place. The small, furry 'warm blooded' scampering things - a minor branch from some obscure reptile that had been so insignificant for over 100 millions years - rose to the challenge.

It was a time of change. Flowering plants had appeared and provided a new rich food source. Although some mammals still felt compelled to lay eggs, they were in the minority, and most mammals became parents to live, young miniature adults which had to be lovingly reared. The mammals split into three distinct groups, members of which still frequent the Earth in modern times. The most primitive group, the monotremes such as echidnas and platypuses, still lay eggs - a throw-back to their reptile past - but, unlike reptiles, the young feed on milk produced by the mother after hatching. The second group are the marsupials, such as kangaroos. The young develop inside the mother for a relatively short period before emerging and are then kept safely inside a pouch and fed on milk while they grow. But the most common and successful group are the placentals, and include man himself. The young are kept inside the mother's body and are nourished by the blood supply of the placenta while they develop, becoming a complex miniature adult in the protected environment of the womb, before being born. They then take milk from their mothers and are reared, often in an advanced social society.

As with the reptiles, some mammals became frustrated with the limitations of the land and decided to take advantage of the air and sea.

Bats, for example, evolved around 50 million years ago, probably beginning as gliders with membranes between their 'fingers' to help free-falling. Sea-dwelling forms such as whales and dolphins developed thick layers of fat instead of fur for insulation, and specialised swimming gear.

Despite the incredible nature of creatures that the Earth had already seen, it was in for an even greater shock. The greatest manifestation was yet to come. Within the past few million years or so, there has come about on Earth a type of creature even more influential than the dinosaurs. The success of this species has demanded a supreme intelligence and an exacting physical ability and manual dexterity. It has developed advanced communications through oral and written language, an unsurpassed ability to travel, and complex societies capable of producing new generations who could build on previously acquired knowledge and ideas. This new species we call mankind.

The presence of man has affected every other living creature on Earth - such is our power - and yet we have really only just arrived. The artists for evolutionary textbooks usually have a field day when they try and depict the gradual transformation of a 'typical' man from a chimpanzee-like primate. The series of images almost always begins with a hairy ape walking on all fours and the following intermediates slowly rise to a standing position. They also suffer an increasing degree of hair loss and gradually develop more handsome, rather chiselled, features. In truth, this single clear pathway from an ape to a human being stems more from the imaginative skills of the creative designer than from known facts.

There is much debate of, not only the what, but also the where and when of human origination. Many have dedicated their lives to archaeological digs, steadfastly searching for 'The Missing Link' between apes and humans, or even just a part of this puzzling jigsaw. The fragmentary pieces of skull and bone that have been so carefully recovered after years of searching, have only added to a greater confusion. It would appear that there are several jigsaws, in fact, and they are all double-sided!

The sequence is patchy and in dispute, but generally starts with *Australopithecus*, a humble primate of about 4 to 5 millions years ago with a small head and large teeth. Then there are *Homo habilis* and *Homo rudolfensis* who supposedly developed into *Homo erectus* and *Homo ergaster*, larger-brained and more human-like hominoids of less than 2 million years ago. Then there was a branch off into a puzzling group called the Neanderthals - quite intelligent and more human-like creatures. And ultimately a predecessor called Cro-Magnon Man appeared - a very recent arrival on the scene at approximately 15,000 to 20,000 years ago. This fellow was a master not only of the useful tool and even language, but also a connoisseur of art and jewellery. Nobody is agreed on where each new form begins and ends. And nobody has the evidence to confirm that this is indeed a true description of the development of *Homo sapiens* - our very own selves.

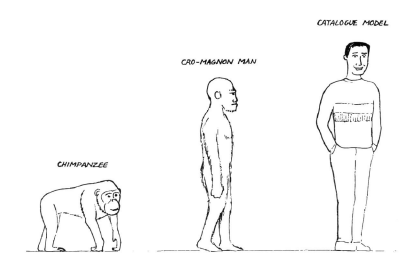

My Family and Other Animals

Less than 200 years ago, mankind had an unchallenged view of his existence. The age of the Earth was estimated to be tens of thousands of years old - rather than thousands of millions. And the overriding belief was that all of the different creatures on Earth, including man, had been individually created and placed here by an omnipotent being, a creator of the universe. Mythologies from all over the world tell of one deity or another who is supposed to be responsible for all of this. In fact, it had never even crossed people's minds that millions of different creatures had wandered about in past ages, or that there had been a gradual change in these animals and plants over time. And if anyone had suggested that human beings were distant relations of these animals, descended from an ape-like creature in a 'remote' continent, they would probably have been called insane.

When a man named Charles Darwin published *The Origin of Species by Means of Natural Selection*, in 1859, these beliefs were irrevocably challenged. Darwin's book was originally intended as an introductory text. He was a keen naturalist and had previously travelled around South America making a number of very detailed observations of many different plants and animals. On his return, he spent years documenting his extensive findings, and eventually his work led him to formulate a number of ideas on how animals had changed over time. The book that he finally published contained only a fragment of the evidence that he had accumulated but, so detailed and lucid was it, that the scientific community was taken by storm. At the time there was huge public debate and much heated argument, but his general idea gradually gained more and more credibility over time. Nowadays, his theory is considered by many to be a highly reasonable explanation of how all life on Earth has evolved.

The theory of evolution by natural selection begins with a very simple observation. The number of offspring that are born or produced, be they tadpoles or seeds, most often exceeds the number that survive. There is a daily battle that must be fought by all animals and plants - against each other for food and mates, against the climate, and against predators and disease. It is a continual lifelong struggle that every individual faces - not only to ensure its own continued survival but also that of the species it represents. Only those plants and animals that are most able to live and reproduce in the face of a vast assortment of challenges that life poses, can survive - at the expense of others that are less well suited. This is the concept of 'survival of the fittest'.

From the evidence he gathered during his travels and from long years of study, Darwin concluded that creatures were not of a fixed design, but had gradually arisen from different ancestors, and that a long line of changing forms had led to the creatures we recognise today. He proposed that the main mechanism for this evolving process involved nature herself, and termed it 'natural selection'. During the period when Darwin was formulating his ideas, nobody knew anything about genes or DNA and the changes they may bring.

Nonetheless, Darwin suggested that an organism would occasionally be born with a new characteristic which, by chance, would be advantageous for that individual. It could be faster legs with which to run away from a predator, thicker fur to help it survive the winter, or a sharper beak to crack seeds more efficiently with. Creatures or plants that had a new adaptation, he proposed, were better able to survive than other members of the species. Gradually, the characteristic would be passed down throughout successive generations because of the advantage it gave to those individuals which carried it. Ultimately, the entire species would have the adaptation, in response to the selective forces of nature.

There are two crucial points in this scenario. Firstly, the small, advantageous adaptation that occurred in the first individual did so purely by chance. It was not designed by the hand of a supreme being. This has the knock-on implication that an adaptation cannot be 'willed' into effect by an animal just because it is needed. If that were the case, no species would have ever become extinct. Secondly, the new adaptation gradually worked its way into the entire species as a result of the 'blind' forces of natural selection, not by the direction of a supreme being.

Darwin's theory, therefore, ruled out the need for religion. And this is probably the main reason why the tremendous furore he created back in the 1800s is still going today.

This, coupled with the peculiar antagonism which *neo-Darwinists* seem to provoke.

To some people, neo-Darwinism sounds like a dodgy political movement which one would be well advised to steer clear of. In fact, it is the term used to describe the coupling of Darwin's theory and our new found knowledge of genetics. We now know that our genes, made up of DNA, can spontaneously change very slightly, or *mutate*. The cause may be a toxic chemical or radiation for example, or it may simply be a chance accident in one of our many millions of cells. The effects of genetic mutation vary tremendously. Depending on the gene that is affected, the individual may develop some crippling disease or abnormality, or there may be absolutely no noticeable outward change at all. In a very, very small number of cases, the mutation may result in a beneficial effect. This is the vital mechanism which lends so much support for Darwin's hypothesis. A small genetic change can bring about a subtle, but critical, adaptation in an offspring. The new and improved characteristic can then offer the creature an advantage over its brothers and sisters in the cruel hostile world. Ultimately, the entire species will sport it.

Given a few billion or so years, the neo-Darwinists say, these cumulative progressive changes have not only created some pretty nifty characteristics within a species, the eye or the brain for example, but they have also conspired to produce the diverse range of plant and animal species that occupy every inhospitable region of the Earth. And mankind.

The Evidence

The first piece of hard factual evidence for evolution really is hard and factual. The fossil record is the only description we have of the plants and animals that have existed on Earth in bygone eras. It shows us that many different types of organisms have indeed lived at some stage in the distant reaches of history, and it also demonstrates that the flora and fauna have changed over time. Although it is notoriously incomplete - creatures appear and disappear in fits and starts, there are few obvious connections between them, and there are tantalising gaps just when things get interesting - it is still the most powerful piece of evidence for the notion that an evolution of some sort has occurred. What must be remembered, however, is that it does not explain how or why organisms have changed, and neither does it lend a huge amount of support for the mechanism that is proposed for the theory of evolution - a fact admitted by Darwin himself.

The proponents of the theory would have a much better case on their hands if the fossil record was a little more accommodating - illustrating their case with a few well-defined evolutionary pathways, for example. The trouble is that fossilisation is a very random and difficult business and this is often used as the main argument for why the fossil record is so incomplete. For those who wish to become preserved for thousands of years as a traditional fossil, dying in or near a large body of water is usually a prerequisite. Over time, you are gently washed with fine sediment which slowly but surely turns to rock. Hard parts of the body, such as bones and teeth, become replaced by minerals and are essentially part of the rock, but there remains a subtle impression, a skeletal daguerreotype of the creature you once were.

To conveniently die near a river, not be eaten by predators, and lie undisturbed for approximately 10,000 years whilst being immortalised in rock, is quite a tall order. Not surprisingly, perhaps, the fossil record is not so much of a record as a collection of snapshots of the past.

The fact that there is little fossil evidence of a step-by-step evolution of creatures and plants is a thorny problem, but yet another difficulty is being able to date the fossilised remains that are found. Accurately judging the age of fossils is essential in order to place a creature in the evolutionary tree, but standard techniques are not infallible. Radioactive dating has become a standard method of determining the age of rock and, by inference, the age of fossils preserved in or near it. The potassium-argon or K/Ar method, for example, measures the amount of change in radioactive potassium to argon. It can only be used accurately in the analysis of rocks containing radioactive material, i.e. volcanic or igneous rock, and is therefore unsuitable for assessing sedimentary rock - the type that fossils are mainly found in. In order to place sedimentary rock in a certain point in history, and hence the fossils it contains, the igneous rock surrounding it or intruding into it is dated. The dating may then be extrapolated even further. These fossils can be compared to those found elsewhere in the world, where there is no volcanic rock. If they are 'similar' enough, these fossils may be attributed with the same age.

Radiocarbon dating is a more commonly known method of dating, not for fossils but rather plant or animal remains and artifacts, up to about 50,000 years old. All living things contain carbon, mainly in the form of carbon 12, but there is also a form of radioactive carbon, carbon 14, which is produced in the atmosphere by the action of the sun. This is absorbed into the bodies of plants and animals throughout their lives and so a small proportion of the carbon in all tissues is carbon 14. After death, carbon 14 decays into nitrogen at a fixed rate. Thus, by measuring the amount of carbon 14 in a sample of bone or wood for example, it is possible to calculate how old it is. This method of dating has its limitations as well, however. Samples may be contaminated by contact with other carbon-containing (i.e. organic) material - even microscopic dead skin cells from someone who has handled the sample. And there is a lack of certainty concerning whether we really do know the correct speed at which carbon 14 decays - volcanic eruptions disrupt the atmosphere and may make the material appear older, and we cannot be sure that it has always been decaying constantly and at the rate we have defined today.

Even being able to date a fossil with any degree of accuracy doesn't solve all the problems. It can still only tell us a limited amount of information, perhaps the creature had only just developed into that form, or perhaps it was a member of a species that had already been in existence for tens of thousands of years. It may have had related species of which no members were preserved. Or it may have become extinct shortly after, thus becoming a dead-end in that particular line of creatures.

Structure and Form

The case for evolution often draws upon a second line of evidence for its defence. As well as the fossil record, which gives us a glimpse into the past, 'informed' guesses about evolution are made from animals and plants that are around today. A diet or even small dose of *Doctor Who* and *Star Trek* prepares us for alien life forms which range from abstract shimmering entities of energy, to humanoid gentlemen in close-fitting silver boiler suits. But, however close our appearances, we are never led to believe that they could possibly share a common ancestry with us.

In contrast, the hugely different creatures alive on Earth today are often closely compared, for evidence to suggest that we really all are one big extended family. In fact, there are many structures that seem to stick to a certain pattern. One of the main pieces of evidence for evolution, which suggests that mammals, reptiles, amphibians and birds have all descended ultimately from one ancestor, comes from this type of comparison, called 'structural homology'. And it is true that most creatures on Earth are of a standard design - two eyes, a nose and mouth, a heart and brain, blood and nerves... Researchers can also analyse the fine bone structure of comparable body parts. For example, looking at the typical forelimbs of a huge variety animals it is possible to see that the bones share striking similarities even though each has become adapted for a particular use - the wing for flying, the flipper for swimming or the arm for waving, for example.

The study of embryology also points in a similar direction. Even animals that do not possess outwardly similar characteristics often do share similarities at one stage in their embryonic development. During the development of many animals in the womb, transient structures appear which do not have any place in the recognisable creatures that eventually emerge. Embryos of vertebrates, for example, go through a phase where they exhibit gill slits. The evidence suggests that all vertebrates, including humans, are descended from common fish ancestors which possessed gill slits.

The theme can be extended even further to a molecular level, to proteins - the molecules that are made up of building blocks called amino acids. Of the hundreds of amino acids that exist, only 20 are used by all animals to make their proteins. Some proteins are common to many kinds of animals and plants but there are slight variations in the amino acid sequence. By comparing the same protein from different species, the number of differences can be used to assess the 'closeness' of two different animals. For example, the protein known as 'cytochrome c' is found in the majority of cells in many animals and plants and is used in the manufacture of energy. The cytochrome c of humans and chimpanzees consists of exactly the same sequence of 104 amino acids. The sequence differs to that of rhesus monkeys by one amino acid and to that of horses by 10 amino acids. This suggests that we are very closely linked to chimpanzees in the evolutionary tree, slightly less closely linked to rhesus monkeys, and even less closely linked to horses!

It is, however, the actual structure of life itself - DNA - which provides the most compelling evidence yet that every organism on this planet is related in some way. The DNA molecule takes exactly the same form, whatever plant or animal it appears in. The twisted 'rope-ladder' design features 'rungs' (or bases) all the way along its length - millions and millions - which form different genes, chromosomes and ultimately life forms. Throughout the whole of nature, the DNA of each and every chromosome contains only four kinds of bases. The only difference between organisms is the sequences of these bases, which then determine such diverse life forms as a mouse and an elephant, a blade of grass and an oak tree. But from bacteria to man, the ultimate life force - the DNA - is constant.

DNA evidence also suggests very strongly that mankind is indeed closely related to our ape-like companions. By comparing the DNA of each species, geneticists can quantify how much genetic material each animal has in common with another. Over 95% of the DNA found in human beings is the same as that found in chimpanzees and gorillas. The differences that exist between us and them are contained in just 3% of our DNA. Does this mean that humans, gorillas and chimpanzees are distant cousins, that we all have a common primate ancestor?

The Truth

There can be no doubt that plants and animals have evolved over time. And even that man himself has developed from the same building blocks of life that are present in all other animals and plants. It is clear that some genetic engineering on a stupendous scale has been going on. Even mankind, with his new biological technology, cannot compare with this genetic manipulation - the creation of millions of completely new life forms over billions of years. The real question is *how* has this genetic engineering taken place?

Does the evidence point to evolution as a blind, natural force that has simply built upon our DNA's natural tendency to mutate? Or does it convince us that evolution was by careful design, that there really is an incredibly intelligent deity out there who knows a thing or two about molecular biology?

The truth is that you can read the evidence either way. If you want to believe in an almighty creator then the facts fit just as easily as if you want to believe in a theory of evolution which is driven entirely by the forces of natural selection. There is simply no conclusive proof one way or the other. Neo-Darwinists may argue about microevolutionary changes which occur in a limited setting - the changing colours of a moth perhaps in response to industrial pollution. But to conduct an experiment to test the theory, to show 'random' evolution of an entirely new form of creature would take thousands of years and an unfeasibly large grant.

Yes, the theory of evolution by nature is a neat explanation, but *no,* it is not necessarily more plausible than the theory of evolution by design, simply because it is based on 'science'. Ultimately, we must remember that our imaginations, and hence our concepts of science, are limited.

And we must remain humbled by whatever genetic engineer has created us.

EPILOGUE

Many years ago when I was a young student, biology was my favourite subject at school. Admittedly, there were the long dreary afternoons when we spent hours peering down microscopes, trying (and usually failing) to match up mangled plant remains with perfect textbook pictures. But my main memories are of the fun-filled field trips where seaweed became associated with limitless mirth, and sand dunes with side-splitting frivolity. Mention the topic of genetics, however, and I would instantly switch off as my mind conjured up grey images of boring, incomprehensible television programmes on BBC2. The main character of these was invariably an earnest, slightly scruffy-looking man in a brown pullover[1] with out-of-control hair and a beard - the archetypal '70s scientist. He was most often to be found standing in front of a blackboard on which were scribbled peculiar symbols, or next to a multicoloured plastic model of strange design. In either case I would feel instantly compelled to leap forward and change the channel. (No remote control in those days!)

Several years on and I still cringe at the sight of a complex yellow and green plastic model, but my interest in the shabby scientist and the subject over which he enthused has grown ten-fold. This change in attitude has certainly not arisen from the numerous dry textbooks I was forced to plough through, but rather from a gradual realisation of the implications of genetics in our lives - our past, present and future.

---

[1] No disrespect intended here. It is a sad fact that brown pullovers, particularly those of the V-necked and tank-top variety, have had a lot of bad press over the years through absolutely no fault of their own.

Attack of the Genetically Engineered Tomatoes

The study of DNA has helped elucidate who we are, in a biological sense, and how we continue to exist. It has enabled us to manipulate the living essence of other animals and plants for our own ends. And it has had a huge effect on the way society perceives the history of the human race, and indeed the evolution of all life on Earth.

Most people actually know more about genetics than they think. We all automatically *know* that children will not look exactly like either parent, but they will have characteristics from both the mother and father, and possibly even the grandparents. This is because we inherit half of our genetic material from each parent. Similarly, when we look at a brother and sister, we're not surprised to find that they are not identical to each other, even though they come from exactly the same parents. This is because the selection of genetic material that each parent passes on is different in every case. Thirdly, we know instinctively that it is impossible for one type of animal to breed with another. We are all aware of the fact that the genetic material of two different species is so incompatible that it precludes successful mating.

BOB KNEW EXACTLY WHAT WAS HAPPENING, HIS FAMILY HAD A HISTORY OF STREAKING DISORDERS.

These basic concepts that shape our lives so fundamentally are at first glance just some of the unalterable rules of nature that we have lived with for centuries, and will surely continue to live with for the rest of human existence. And yet are they? We have seen in previous chapters how easily genes can now be transferred from plant to animal and even from man to mouse. The genetic manipulation of the human race is just a stone's throw away.

Our Future in Their Hands

On Wednesday 10th August 1994, the *Daily Express* newspaper published a fantastical article under the heading of 'Medical Report'. The sensational headline screamed *'Now you can be sure how long you have left to live'* and the following text contained a selection of exquisite phrases such as *'Genes science will map out the whole course of your life'* and *'Each morning they wake up, the first thing our grandchildren will do is to check how many days they have left to live'*. Referring to parents of the future, we hear that *'In the garden, their regulation single child ... runs among the genetically-manipulated flowers and trees. His deeply-tanned skin, hazel eyes and dark hair all chosen before he was even conceived'*.

If we let our imaginations run wild, we can imagine a world where all of our genes have been de-coded, allowing manipulation of any part of our bodies. Media exaggeration aside, however, we are far from understanding all of our genes. We do know that they determine characteristics such as eye and skin colour, height and nose size, for example, but in a very complex fashion; there is no single gene which can be pinpointed and labelled as the gene for blue eyes. They can also give us clues of genetic disorders we may be prone to, but they will never tell us exactly how long we have to live. The real danger is less conspicuous, and is already upon us.

The fantasies that were so adeptly presented in the *Daily Express*'s imaginary scenario would once have been more at home in a *Star Trek* movie - Captain Kirk would have been valiantly espousing the ethical objections and chastising such a morally repugnant alien society. But this situation is closer to home than we would like to think. Our growing knowledge of genes and their effects has already led to frightening developments. In societies where male babies are more 'desirable' than females, abortions have been used to destroy unwanted fetuses. In China, doctors are legally compelled to recommend abortion for any fetus where a prenatal test has found a serious genetic disorder. And, in 1995, a chilling new law came into effect there which has raised concern all over the world. All individuals who plan to wed must now undergo testing to find out whether they are carrying 'unwanted' genes for diseases that could be passed on. For those unfortunates who find themselves as carriers, marriage is only permitted if they agree never to have children, and if they consent to sterilisation or long term contraception.

We must ask ourselves, how far reaching are the implications of the law in China? If a society is ruthless enough to restrict the freedom and rights of individuals to marry and even bear children, there will surely be no end to the evil that could ensue. What exactly constitutes a genetic disorder anyway? In recent years scientists have located genes for violent, aggressive tendencies, obesity and depression. Some researchers have even pointed to genes for homosexuality. If we begin to limit the human race on anything, where will it end? Nobody *is* perfect and the idea of *perfection* is abhorrent anyway. That's not to say that painful and distressing genetic disorders shouldn't be fixed if we have the technology, but the wider issue demands the acceptance of human life as it is.

This is a crucial time for the human race. We must all sit up and take notice of genetics research, and it is no excuse to claim ignorance of science. You don't have to be able to grasp the complexities of DNA and chromosomes in order to take a stance. It is the moral duty of each of us to be aware of the fundamental concepts which affect us all.

And to have an *opinion*.

BIBLIOGRAPHY

Space does not allow for a definitive bibliography for the subject matter contained in this book. The references listed below are a selection of publications which may be of interest to the reader, and which were of particular value in the compilation of this work.

INTRODUCTION / CHAPTER 1

Baker RJ, Van Den Bussche RA, Wright AJ, Wiggins LE, Hamilton MJ, Reat EP, et al. High levels of genetic change in rodents of Chernobyl. Nature 1996;380:707-8.

Brunner HG. MAOA deficiency and abnormal behaviour: perspectives on an association. Ciba Found Symp 1996;194:155-64.

Devlin B, Daniels M, Roeder K. The heritability of IQ. Nature 1997;388:468-71.

Dubrova YE, Nesterov VN, Krouchinsky NG, Ostapenko VA, Neumann R, Neil DL, et al. Human minisatellite mutation rate after the Chernobyl accident. Nature 1996;380:683-6.

Harrington JJ, Van Bokkelen G, Mays RW, Gustashaw K, Willard HF. Formation of de novo centromeres and construction of first-generation human artificial microchromosomes. Nat Genet 1997;15:345-55.

Jordan BD, Relkin NR, Ravdin LD, Jacobs AR, Bennett A, Gandy S. Apolipoprotein E epsilon4 associated with chronic traumatic brain injury in boxing. JAMA 1997;278:136-40.

Keverne EB, Fundele R, Narasimha M, Barton SC, Surani MA. Genomic imprinting and the differential roles of parental genomes in brain development. Brain Res Dev Brain Res 1996;92:91-100.

Koopman P, Gubbay J, Vivian N, Goodfellow P, Lovell-Badge R. Male development of chromosomally female mice transgenic for *Sry*. Nature 1991;351:117-21.

Koopman P, Munsterberg A, Capel B, Vivian N, Lovell-Badge R. Expression of a candidate sex-determining gene during mouse testis differentiation. Nature 1990;348:450-2.

Lenders JW, Eisenhofer G, Abeling NG, Berger W, Murphy DL, Konings CH, et al. Specific genetic deficiencies of the A and B isoenzymes of monoamine oxidase are characterized by distinct neurochemical and clinical phenotypes. J Clin Invest 1996;97:1010-9.

Li L, Hoffman RM. The feasibility of targeted selective gene therapy of the hair follicle. Nat Med 1995;1:705-6.

Li L, Hoffman RM. Topical liposome delivery of molecules to hair follicles in mice. J Dermatol Sci 1997;14:101-8.

McClearn GE, Johansson B, Berg S, Pedersen NL, Ahern F, Petrill SA, et al. Substantial genetic influence on cognitive abilities in twins 80 or more years old. Science 1997;276:1560-3.

McGuffin P, Thapar A. Genetic basis of bad behaviour in adolescents. Lancet 1997;350: 411-2.

Nelson RJ, Demas GE, Huang PL, Fishman MC, Dawson VL, Dawson TM, et al. Behavioural abnormalities in male mice lacking neuronal nitric oxide synthase. Nature 1995;378:383-6.

Pearce F. Devil in diesel. New Sci 1997;156(2105):4.

Penny GD, Kay GF, Sheardown SA, Rastan S, Brockdorff N. Requirement for Xist in X chromosome inactivation. Nature 1996;379:131-7.

Plomin R, McClearn GE, Smith DL, Vignetti S, Chorney MJ, Chorney K, et al. DNA markers associated with high versus low IQ: the IQ Quantitative Trait Loci (QTL) Project. Behav Genet 1994;24:107-18.

Plomin R, Pedersen NL, Lichtenstein P, McClearn GE. Variability and stability in cognitive abilities are largely genetic later in life. Behav Genet 1994;24:207-15.

Turner G. Intelligence and the X chromosome. Lancet 1996;347:1814-5.

Wiley LM, Baulch JE, Raabe OG, Straume T. Impaired cell proliferation in mice that persists across at least two generations after paternal irradiation. Radiat Res 1997;148:145-51.

CHAPTER 2

Balding DJ, Donnelly P. How convincing is DNA evidence? Nature 1994;368:285-6.

Cano RJ, Poinar HN, Pieniazek NJ, Acra A, Poinar GO, Jr. Amplification and sequencing of DNA from a 120-135-million-year-old weevil. Nature 1993;363:536-8.

Charatan FB. Peruvian mummy shows that TB preceded Columbus. BMJ 1994;308:808.

Coghlan A. Iceman's relatives traced to northern Europe. New Sci 1994;142(1931):6-7.

Dickman S. A real culture shock. New Sci 1997;155(2091):4-5.

DNA clue. New Sci 1993;138(1878):12.

DNA 'trapped badger killers'. Daily Mail 1997 Sep 19:17.

Horgan J. High profile. Scientific American 1994;271:33-6.

Housman DE. DNA on trial - the molecular basis of DNA fingerprinting. N Engl J Med 1995;332:534-5.

Lewin R. Fact, fiction and fossil DNA. New Sci 1994;141(1910):38-41.

MacKenzie D. Endangered or minke, sir? New Sci 1997;155(2097):14.

Pääbo S. Ancient DNA. Scientific American 1993;269:60-6.

Perlman SE. Nicole's blood on Bronco rug? Sources say DNA tests positive. The Denver Post 1994 Sep 28:2A.

Quirke S, Spencer J, editors. The British Museum book of ancient Egypt. London: British Museum Press, 1992.

Ranson P. Consent: the legal aspects. Good Clinical Practice Journal 1994;1:24-5.

Salo WL, Aufderheide AC, Buikstra J, Holcomb TA. Identification of *Mycobacterium tuberculosis* DNA in a pre-Columbian Peruvian mummy. Proc Natl Acad Sci USA 1994;91:2091-4.

Smith GE, Dawson WR. Egyptian mummies. London: Kegan Paul International Ltd., First publ. 1924, Republished 1991.

Spigelman M. Studying ancient DNA. BMJ 1994;308:1370.

Spigelman M, Lemma E. The use of the polymerase chain reaction (PCR) to detect *Mycobacterium tuberculosis* in ancient skeletons. Int J Osteoarch 1993;3:137-43.

Spindler K. The man in the ice. London: Weidenfeld and Nicolson, 1994.

Stewart SA. Credibility 'low' for DNA tests. USA TODAY 1994 Sep 23:4A.

Szabo M. DNA test traps whale traders. New Sci 1994;142(May 28):4-5.

Wilkie T. Gene hunters. London: Channel 4 Television, 1995.

CHAPTER 3

Advisory Committee on Novel Foods and Processes. Report on the ethics of genetically modified foods. London: HMSO, 1993.

Barefoot SF, Beachy RN, Lilburn MS, Goldhammer AR, Harlander SK, Sullivan SL. Labeling of food-plant biotechnology products. Issue paper - Council for Agricultural Science and Technology, 1994 No.4:1-8.

British Medical Association. Our genetic future: the science and ethics of genetic technology. Oxford: Oxford University Press, 1992.

Coghlan A. Genetic gun makes rice growers' day. New Sci 1991;(Nov 2):23.

Day M. Superbug spectre haunts Japan. New Sci 1997;154(2080):5.

Department of the Environment. Fast track for low risk releases of genetically modified organisms. London: Department of the Environment,1994.

Dixon B. Biotech consumers discriminate. Bio/Technology 1995;13:20-1.

Gershon D. Genetically engineered foods get green light. Nature 1992;357:352.

Gledhill M, McGrath P. Call for a spin doctor. New Sci 1997;156(2106):4-5.

Haq TA, Mason HS, Clements JD, Arntzen CJ. Oral immunization with a recombinant bacterial antigen produced in transgenic plants. Science 1995;268:714-6.

Khan MR, Ceriotti A, Tabe L, Aryan A, McNabb W, Moore A, et al. Accumulation of a sulphur-rich seed albumin from sunflower in the leaves of transgenic subterranean clover (*Trifolium subterraneum* L.). Transgenic Res 1996;5:179-85.

Kiernan V. Chicago chokes on 'altered' pizza. New Sci 1993;(Sep 4):5.

Kiernan V. Yes, we have vaccinating bananas. New Sci 1996;151(2048):6.

Kleiner K. Fields of genes. New Sci 1997;155(2095):4.

Klinger T, Elam DR, Ellstrand NC. Radish as a model system for the study of engineered gene escape rates via crop-weed mating. Conservation Biology 1991;5:531-5.

Kondo K, Miura Y, Sone H, Kobayashi K, Iijima H. High-level expression of a sweet protein, monellin, in the food yeast *Candida utilis*. Nat Biotechnol 1997;15:453-7.

Ma SW, Zhao DL, Yin ZQ, Mukherjee R, Singh B, Qin HY, et al. Transgenic plants expressing autoantigens fed to mice to induce oral immune tolerance. Nat Med 1997;3: 793-6.

Martin S, Tait J. Release of genetically modified organisms: public attitudes and understanding 'summary report'. The Open University, Centre for Technology Strategy, 1992.

Pincock S. Spud gun targets disease. New Sci 1997;155(2092):18.

Pinholster G. Debatable edibles: bioengineered foods. Environmental Health Perspectives 1994;102:636-9.

CHAPTER 4

Campbell KH, McWhir J, Ritchie WA, Wilmut I. Sheep cloned by nuclear transfer from a cultured cell line. Nature 1996;380:64-6.

Concar D. The organ factory of the future? New Sci 1994;142(1930):24-9.

Dorling A, Riesbeck K, Warrens A, Lechler R. Clinical xenotransplantation of solid organs. Lancet 1997;349:867-71.

Evans C, Houston S. Monsters or a miracle? Daily Mail 1996 Mar 7:1,6-7.

Galimand M, Guiyoule A, Gerbaud G, Rasoamanana B, Chanteau S, Carniel E, et al. Multidrug resistance in *Yersinia pestis* mediated by a transferable plasmid. N Engl J Med 1997;337:677-80.

House of Commons Select Committee on European Legislation. 19th Report. Session 1992-93. Patenting of Biotechnological Inventions (4148/93). London: HMSO, 1993.

Kleiner K. US bans 'hormone free' milk label. New Sci 1994;141(1914):5.

MacKenzie D. Can we make supersalmon safe? New Sci 1996;149(2014):14-5.

MacKenzie D. Patents on life sneak through back door. New Sci 1994;142(1925):6-7.

Majzoub JA, Muglia LJ. Knockout mice. N Engl J Med 1996;334:904-7.

Nuffield Council on Bioethics. Animal to human transplants: the ethics of xenotransplantation. London: Nuffield Council of Bioethics, 1996.

Palmiter RD, Brinster RL, Hammer RE, Trumbauer ME, Rosenfeld MG, Birnberg NC, et al. Dramatic growth of mice that develop from eggs microinjected with metallothionein-growth hormone fusion genes. Nature 1982;300:611-5.

Preston WP, McGuirk AM, Jones GM, Caswell JA. Consumer reaction to the introduction of bovine somatotrophin. In: Caswell JA, editor. Economies of Food Safety. New York: Elsevier Science Publishing Co., Inc., 1991:189-210.

Shuldiner AR. Transgenic animals. N Engl J Med 1996;334:653-5.

The Advisory Group on the Ethics of Xenotransplantation. Animal tissue into humans. London:Department of Health, 1997.

Wallach P, Mukerjee M, Kumar S. Regulating the body business. Scientific American 1996;274:8-10.

Wilmut I, Schnieke AE, McWhir J, Kind AJ, Campbell KH. Viable offspring derived from fetal and adult mammalian cells (published erratum appears in Nature 1997;386:200). Nature 1997;385:810-3.

Wise J. New authority to monitor xenotransplantation experiments. BMJ 1997;314:247.

Wise J. Pig virus transfer threatens xenotransplantation. BMJ 1997;314:623.

CHAPTER 5

Alper JS. Genetic complexity in single gene diseases. BMJ 1996;312:196-7.

Axworthy D, Brock DJ, Bobrow M, Marteau TM. Psychological impact of population-based carrier testing for cystic fibrosis: 3-year follow-up. Lancet 1996;347:1443-6.

Baird R. It's none of their business. New Sci 1997;156(2102):13.

Blau HM, Springer ML. Gene therapy - a novel form of drug delivery. N Engl J Med 1995;333:1204-7.

Brock DJ. Prenatal screening for cystic fibrosis: 5 years' experience reviewed. Lancet 1996;347:148-50.

Brunner EJ, Sheppard J, Ravetz J. Public is concerned about gene testing [letter]. BMJ 1997;314:1552-3.

Davis PB. Cystic fibrosis: new perceptions, new strategies. Hosp Pract 1992;27:79-118.

Dubowitz V. The muscular dystrophies - clarity or chaos? N Engl J Med 1997;336:650-1.

Gene therapy for CF gets go-ahead. Marketletter 1992;19(49):20.

Hyde SC, Gill DR, Higgins CF, Trezise AE, MacVinish LJ, Cuthbert AW, et al. Correction of the ion transport defect in cystic fibrosis transgenic mice by gene therapy. Nature 1993;362:250-5.

Kmietowicz Z. Health put at risk by insurers' demands for gene test results. BMJ 1997;314:625.

Larson JE, Morrow SL, Happel L, Sharp JF, Cohen JC. Reversal of cystic fibrosis phenotype in mice by gene therapy in utero. Lancet 1997;349:619-20.

Livingstone J, Axton RA, Gilfillan A, Mennie M, Compton M, Liston WA, et al. Antenatal screening for cystic fibrosis: a trial of the couple model BMJ 1994;308:1459-62.

Motluk A. Government does U-turn over genetics watchdog. New Sci 1996;149(2021):10.

Report of the Committee on the Ethics of Gene Therapy. London, 1993.

Shaw D. When DNA turns traitor. New Sci 1995;(Mar 25):28-33.

Smeets HJ, Nillesen WM, Los F, Busch HF, Korneluk RG, Wieringa B, et al. Prenatal diagnosis of myotonic dystrophy by direct mutation analysis [letter]. Lancet 1992;340: 237-8.

Stern RC. The diagnosis of cystic fibrosis. N Engl J Med 1997;336:487-91.

Super M, Schwarz MJ, Malone G, Roberts T, Haworth A, Dermody G. Active cascade testing for carriers of cystic fibrosis gene. BMJ 1994;308:1462-5.

The Cystic Fibrosis Genotype-Phenotype Consortium. New Engl J Med 1993;329:1308-9.

UK gene therapy trial approved. Marketletter 1993;20(5):23.

Warden J. Britain creates watchdog for genetic tests. BMJ 1996;312:141-2.

Warden J. MPs seek more control over genetics. BMJ 1996;312:1119.

Wilson JM. Adenoviruses as gene-delivery vehicles. N Engl J Med 1996;334:1185-7.

Wilson JM. Vehicles for gene therapy. Nature 1993;365:691-2.

Yates S. Staying in the fast track. In: Muscular Dystrophy Group. Annual review 1993. London: Muscular Dystrophy Group of Great Britain & Northern Ireland, 1993:9-16.

CHAPTER 6

Asvall JE. The WHO wants governments to encourage people to stop smoking [letter]. BMJ 1997;314:1688.

Austoker J. Cancer prevention: setting the scene. BMJ 1994;308:1415-20.

Bakketeig LS, Jacobsen G, Hoffman HJ, Lindmark G, Bergsjo P, Molne K, et al. Pre-pregnancy risk factors of small-for-gestational age births among parous women in Scandinavia. Acta Obstet Gynecol Scand 1993;72:273-9.

Bartecchi CE, MacKenzie TD, Schrier RW. The human costs of tobacco use. N Eng J Med 1994;330:907-12.

Beardsley T. Vital data. Scientific American 1996;274:76-81.

Bischoff JR, Kirn DH, Williams A, Heise C, Horn S, Muna M, et al. An adenovirus mutant that replicates selectively in *p53*-deficient human tumor cells. Science 1996;274:373-6.

Bonn D. Getting under the skin with melanoma vaccines. Lancet 1996;348:396.

Breast Cancer Linkage Consortium. Pathology of familial breast cancer: differences between breast cancers in carriers of *BRCA1* or *BRCA2* mutations and sporadic cases. Lancet 1997;349:1505-10.

Carbone D. Smoking and cancer. Am J Med 1992;93 Suppl 1A:13S-17S.

Charatan FB. Smoker wins damages against US tobacco company. BMJ 1996;313:382.

Coghlan A. Take aim, fire. New Sci 1997;156(2106):28.

Collins FS. *BRCA1* - lots of mutations, lots of dilemmas. N Engl J Med 1996;334:186-8.

Denissenko MF, Pao A, Tang M, Pfeifer GP. Preferential formation of benzo(a)pyrene adducts at lung cancer mutational hotspots in P53. Science 1996;274:430-2.

Doll R, Peto R, Wheatley K, Gray R, Sutherland I. Mortality in relation to smoking: 40 years' observations on male British doctors. BMJ 1994;309:901-11.

Easton DF, Narod SA, Ford D, Steel M. The genetic epidemiology of *BRCA1* [letter]. Lancet 1994;344:761.

Eeles R. Testing for the breast cancer predisposition gene, *BRCA1*. BMJ 1996;313:572-3.

Eng C, Stratton M, Ponder B, Murday V, Easton D, Sacks N, et al. Familial cancer syndromes. Lancet 1994;343:709-13.

Evans DG, Fentiman IS, McPherson K, Asbury D, Ponder BA, Howell A. Familial breast cancer. BMJ 1994;308:183-7.

Ford D, Easton DF, Bishop DT, Narod SA, Goldgar DE. Risks of cancer in *BRCA1*-mutation carriers. Lancet 1994;343:692-5.

Fricker J. Specific *BRCA2* mutations do not occur in all breast-cancer families. Lancet 1996;347:1319.

Gene therapy for melanoma promising. Scrip 1997;(2249):19.

Harris CC, Hollstein M. Clinical implications of the *p53* tumor-suppressor gene. N Engl J Med 1993;329:1318-27.

Healy B. *BRCA* genes - bookmaking, fortunetelling, and medical care [editorial]. N Engl J Med 1997;336:1448-9.

Hecht SS, Carmella SG, Murphy SE, Akerkar S, Brunnemann KD, Hoffmann D. A tobacco-specific lung carcinogen in the urine of men exposed to cigarette smoke. N Engl J Med 1993;329:1543-6.

Josefson D. US doctors warned about test for breast cancer gene. BMJ 1996;312:1057-8.

Kallen K. Maternal smoking during pregnancy and limb reduction malformations in Sweden. Am J Public Health 1997;87:29-32.

Krainer M, Silva-Arrieta S, FitzGerald MG, Shimada A, Ishioka C, Kanamaru R, et al. Differential contributions of *BRCA1* and *BRCA2* to early-onset breast cancer. N Engl J Med 1997;336:1416-21.

Langston AA, Malone KE, Thompson JD, Daling JR, Ostrander EA. *BRCA1* mutations in a population-based sample of young women with breast cancer. N Engl J Med 1996;334:137-42.

Lloyd SA. Stratospheric ozone depletion. Lancet 1993;342:1156-8.

Marks R. Primary prevention of skin cancer. BMJ 1994;309:285-6.

McGregor JM, Young AR. Sunscreens, suntans, and skin cancer. BMJ 1996;312:1621-2.

Nechushtan A, Yarkoni S, Marianovsky I, Lorberboum-Galski H. Adenocarcinoma cells are targeted by the new GnRH-PE66 chimeric toxin through specific gonadotropin-releasing hormone binding sites. J Biol Chem 1997;272:11597-603.

Nguyen DM, Wiehle SA, Koch PE, Branch C, Yen N, Roth JA, et al. Delivery of the *p53* tumor suppressor gene into lung cancer cells by an adenovirus/DNA complex. Cancer Gene Ther 1997;4:191-8.

Peto R. Smoking and death: the past 40 years and the next 40. BMJ 1994;309:937-9.

Phillips AN, Goya Wannamethee S, Walker M, Thomson A, Davey Smith G. Life expectancy in men who have never smoked and those who have smoked continuously: 15 year follow up of large cohort of middle aged British men. BMJ 1996;313:907-8.

Progress in skin cancers. Scrip 1996;(2090/91):25.

Rees JL. The melanoma epidemic: reality and artefact. BMJ 1996;312:137-8.

Rivers JK. Melanoma. Lancet 1996;347:803-6.

Rowe PM. *BRCA1* involved two ways in sporadic cancers. Lancet 1995;345:917.

Sethi T. Lung cancer. BMJ 1997;314:652-5.

Sikora K. Genes, dreams, and cancer. BMJ 1994;308:1217-21.

Skin cancer gene identified. Scrip 1996;(2140):27.

Sporn MB. The war on cancer. Lancet 1996;347:1377-81.

Struewing JP, Tarone RE, Brody LC, Li FP, Boice JD, Jr. *BRCA1* mutations in young women with breast cancer [letter]. Lancet 1996;347:1493.

Thomas F, Zitvogel L. Gene therapy in the treatment of cancer. Scrip Magazine 1997;(Feb):38-43.

Tsukamoto M, Ochiya T, Yoshida S, Sugimura T, Terada M. Gene transfer and expression in progeny after intravenous DNA injection into pregnant mice. Nat Genet 1995;9:243-8.

Tucker JD, Morgan WF, Awa AA, Bauchinger M, Blakey D, Cornforth MN, et al. PAINT: a proposed nomenclature for structural aberrations detected by whole chromosome painting. Mutat Res 1995;347:21-4.

Wakschlag LS, Lahey BB, Loeber R, Green SM, Gordon RA, Leventhal BL. Maternal smoking during pregnancy and the risk of conduct disorder in boys. Arch Gen Psychiatry 1997;54:670-6.

Wright B. Smokers' sperm spell trouble for future generations. New Sci 1993;137(1863):10.

Wright B. Sunscreens and the protection racket. New Sci 1994;141(Jan 22):21-2.

Zhang JF, Hu C, Geng Y, Selm J, Klein SB, Orazi A, et al. Treatment of a human breast cancer xenograft with an adenovirus vector containing an interferon gene results in rapid regression due to viral oncolysis and gene therapy. Proc Natl Acad Sci USA 1996;93: 4513-8.

CHAPTER 7

It is impossible to provide even a selection of references which could begin to cover this subject! However, essential reading for all attempting to grapple with traditional evolutionary theory includes:

Darwin C. The origin of species by means of natural selection. First published John Murray, 1859. Republished by Penguin Classics, 1985.

Dawkins R. The blind watchmaker. London: Penguin Books Ltd., 1988.

Jones S. The language of the genes. Biology, history and the evolutionary future. London: Flamingo, 1993.

Maynard Smith J. Did Darwin get it right? Essays on games, sex and evolution. London: Penguin Books Ltd., 1988.

EPILOGUE

Anderson P. Now you can be sure how long you have left to live. Daily Express 1994 Aug 10:22-31.

Speak, don't hold your peace [editorial]. New Sci 1996;152(2056):3.

Gene warfare - unless we keep our guard up [editorial]. Lancet 1996;348:1183.

INDEX

A

adenovirus 90, 91, 105
Advisory Committee on Genetic Modification 59
Advisory Committee on Genetic Testing 96
Advisory Committee on Novel Foods and Processes 59
Advisory Committee on Releases to the Environment 59
ageing 15
aggression 17
Agrobacterium tumefaciens 52
Alicaligenes eutrophus 53
alpha-1-antitrypsin 71
ampicillin 57
ancient DNA 31, 32
antibiotic resistant bacteria 58
antibiotics 57, 75
apoptosis 20, 83

B

baldness 15
bananas 54
benzo[a]pyrene 102
bovine somatotrophin 75
brain 25
BRCA1 104
BRCA2 104
breast cancer 103
breeding 49
bromoxynil 56
Burgess Shale 113
bystander effect 106

C

calcitonin 71
Calgene 48, 56
Campbell Soup Company 48
Candida utilis 53
carriers 89
cells 23, 26
Cheddar Man 34
cheese production 54
Chernobyl 18
chromosome 11, 27
chymosin 54
cloning 36, 68, 78
collagenase 15
cotton 53, 56
cystic fibrosis 86
cystic fibrosis transmembrane regulator 87
cytochrome c 125

D

Darwin, Charles 120
development 16, 28, 124
diabetes 54
DNA fingerprinting 37
DNA ingestion 58
Dolly 79
Down's syndrome 13

E

ear 24
embryology 124
Enterococcus 58
Escherichia coli 54
Ethical Committee on Genetic Modification of Food 60
European corn borer 57
eye 24

F

Flavr Savr 48, 57
fossils 122

G

GAD 54
Gene Therapy Advisory Committee 95
genome 14
glufosinate 57
Good Clinical Practice 44
granulocyte macrophage-colony stimulating factor 109

H

helicase 15
hepatitis B 54
Homo sapiens 119
Human Genetics Commission 95
Human Genome Diversity Project 43
Human Genome Organisation 45
Human Genome Project 42
human interferon 105

I

imprinted gene 16
insulin 54, 55
insurance 94
intelligence 16, 17

J

junk DNA 43

K

kanamycin 57
Klinefelter's syndrome 13
knock-out animals 77

L

labelling 59
lactoferrin 71
liposomes 91

M

maize 57
melanin 106
melanoma 106
milk production 75
monellin 52
monoamine oxidase A 17
mummy 34
mutagen 18, 85
mutation 18, 84, 100, 102, 104, 121
Mycobacterium tuberculosis 34
myotonic dystrophy protein kinase 93
myotonic muscular dystrophy 92

N

natural selection 120
neo-Darwinism 121
neomycin 57
nerve cell 24
NNK 99
non-melanoma 106
nucleus 26

O

onco-mouse 76
oncogene 76, 102
Ötzi 32
ovarian cancer 104

P

p53 102
passive smoking 99
patents 76
pharming 70
phenylalanine 88
pigs 72
Pikaia 114
pink bollworm caterpillar 56
plasmid 15, 52, 58
plastic 53
Polly 79
polymerase chain reaction 37
potatoes 54, 62
pregnancy 101
proof reading 18

Q

quagga 35

R

rapeseed 61
recombination 16
rice 51

S

Saccharomyces cerevisiae 59
salmon 72
selective breeding 36
sex chromosomes 12
sheep 67, 71
skin cancer 106
smoking 98
SRY 14
Staphylococcus aureus 58
structural homology 124
Sun Protection Factor 107
survival of the fittest 120

T

tobacco 85
tobacco plants 56
tomatoes 48
transgenic organisms 50, 69
transplantation 73
tuberculosis 34
tyrosinase 15

V

vaccination 54

W

Werner's syndrome 15
whale meat 41

Y

Yersinia pestis 58